Modifying Creation?

An Evangelical Alliance Policy Commission Report

Edited by
Donald Bruce and Don Horrocks

Working Group Members

Donald Bruce
John Bryant
Jonathan Burnside
Bob Carling
Peter Carruthers
Avice Hall
Don Horrocks
Tony May

paternoster
press

Copyright © 2001 Evangelical Alliance
Whitefield House, 186 Kennington Park Road, London, SE11 4BT

First published in 2001 by Evangelical Alliance Policy Commission

07 06 05 04 03 02 01   7 6 5 4 3 2 1

Evangelical Alliance Policy Commission is an imprint of
Paternoster Publishing,
PO Box 300, Carlisle, Cumbria, CA3 0QS, UK
and PO Box 1047, Waynesboro, GA 30830-2047, USA

Website: www.paternoster-publishing.com

British Library Cataloguing in Publication Data
A catalogue record for this book is available from the British Library

ISBN 1-84227-100-8

Typeset by WestKey Ltd, Falmouth, Cornwall
Cover Design by Campsie, Glasgow
Printed in Great Britain by
Cox and Wyman Ltd., Cardiff Road, Reading

Contents

About the Policy Commission

The Evangelical Alliance's Millennium Manifesto incorporated a decision to create a Policy Commission, which was initiated in 1999. Its remit was to identify contemporary controversial issues, commission relevant studies, and, in adopting an evangelical viewpoint, recommend appropriate policy statements and affirmations. The issues with which it was primarily to be concerned would typically be of an ethical nature with societal, national or international implications – as distinct from purely theological and doctrinal concerns. Although addressing in the first instance evangelical Christians and churches, its studies are intended to be of wider relevance to the Christian community and society at large as it seeks to offer a co-ordinated response to matters of wider public debate.

The Policy Commission functions as a steering group to the Evangelical Alliance, comprising evangelical representatives from a wide range of academic, scientific and professional disciplines. Its reports are produced following a wide-ranging discussion and consultation process, both internally through the Evangelical Alliance membership, and externally with reference to appropriately qualified academics and practitioners in the relevant fields.

The next report to appear from the Policy Commission will examine the subject of relations between Nation and Faith.

For more information about the Policy Commission write to: Don Horrocks, Manager, Public Affairs, Evangelical Alliance, Whitefield House, 186, Kennington Park Road, London, SE11 4BT.

e–mail:	alilley@eauk.org
Website:	http://www.eauk.org
Telephone:	020 7207 2100
Fax:	020 7207 2150

Foreword

In matters of health and environment, particular issues have the capacity – with a little help from non-governmental organisations – to grip the public mind for a brief moment and then to fade from the limelight, rather like pop songs. For a few weeks or months they are the all-consuming focus of political attention. Then the public focus shifts to some new source of angst. This does not mean that the issue has no real substance or importance. Usually, the questions are important and deserve careful attention and thought. Nor do the issues die completely from the political agenda; they rumble on as part of the established concerns of the broad policy community.

One of the problems of the short concentration spans of the media and their consumers is that the tempos of knowledge are not the same as those of politics in the electronic age. In the heat of the media crisis, policymakers, academics and others initiate a series of studies, designed the better to understand the issues – to answer the questions that expert advisers could not answer in the heat of the moment. For very often, they cannot do so: quite apart from the fact that they and policymakers alike are the prisoners of conventional thinking, experts are constrained by the fact that their information, methodologies, research hardware and even the wisdom of their experience are the product of trying to answer yesterday's questions. This was notably so in the case of BSE. But it is the same in many other cases. Unfortunately, this reality is thoroughly at odds with the public expectation that 'they' should be able to give instant and dependable answers. That they cannot do so contributes to growing public distrust of experts. None of this undermines, however, the great value of studying the issues at leisure, as a basis for future policy.

These general processes can be seen in the case of genetically-modified organisms – GMOs as they are known. Though the technology was already in extensive use, in 1999 in the UK and Europe, GMOs caught public attention. They hit the top of the political charts. Public concern has now shifted to other issues, at least temporarily. But the underlying technical, ethical and, for Christians, theological questions, brought partly and haphazardly into focus in 1999, remain. The fact that public attention is now diverted elsewhere does not mean that these questions are insignificant. They desperately need better illumination for the wider public mind, because the policy issues have not gone away.

In fact, excellent work on the ethical, scientific and policy issues engaged by this form of human intervention had already been done by the spring of 1999, some of it as a matter of fact by Evangelical Christians.* In response to the raising of the political temperature, the UK Government instituted its own series of independent reviews and research. For its part, the Evangelical Alliance's Policy Commission believed that the questions raised by GMO technology were of such importance in themselves, and raised such significant wider issues – including theological issues, that it should say something about them. This book is the product of that work.

---

* M.J. Reiss and R. Straughan, *Improving Nature? The Science and Ethics of Genetic Engineering,* (Cambridge: Cambridge University Press, 1996); D. Bruce and A. Bruce (eds.), *Engineering Genesis,* (London: Earthscan Publications, 1998); ESRC Global Environmental Change Programme (1999); *The Politics of GM Food, Risk, Science and Public Trust,* Special Briefing No. 5, University of Sussex; *Genetically Modified Crops: the ethical and social issues,* Nuffield Council on Bioethics, (May 1999).

In itself, GMO technology is technical, even narrow, in its character. Much public discussion of it, including that in the Christian community, goes on with little accurate knowledge of what is involved. Many people don't know much about it, but they don't like it. The clear explanations given early in this book are welcome and necessary in themselves. But the technology raises a range of much wider questions, going to the heart of:

- The relation between humans and the rest of the created world
- Their right to intervene in and modify that world in their own interests, those of their fellow human beings, and even those of other organisms
- The terms and conditions on which they might do so
- Attitudes to, and management of, risk to humans and other species and eco-systems
- The nature and philosophy of agricultural, forestry and fisheries systems
- The balance to be struck between human development, the eradication of poverty, distribution of wealth between humans, and the protection and enhancement of the environment – and what should be the priorities in the light of the Christian revelation? This is the substance of key debates today about what is known, perhaps unhelpfully, as 'sustainable development'.
- Tensions around the scope and processes of globalisation in today's world, and in particular the role and rights of multinational companies, and of governments and civil society to constrain and regulate them.

And so on.

In short, GMO technology turns out to be a prism through which we as Christians can gain a better understanding of fundamental questions relating to human life, development and the environment. That is why the working group's report, and the Policy Commission's conclusions and recommendations, range so widely. The analysis is thorough, and it and the conclusions deserve very careful consideration by Christian and non-Christian alike – even if they are to be regarded as material for further debate, rather than as a conclusive statement of what ought to be believed and practised without further ado.

The work is also, I believe, further evidence of a wider and very welcome development. That is, the successful marriage in the evangelical camp of detailed understanding of specific policy issues, and the background to them, with serious biblical and ethical understanding, to provide a framework for practical action. At times in the past 30 years, I for one have despaired at the jejune work among evangelicals that has passed for establishing a 'Christian mind' on particular policy issues. This book is far from jejune. It shows that the genre is at last coming of age among evangelicals in the UK. I welcome the fact that the Evangelical Alliance has conceived and nurtured it, as an example of the public voice to which it is aspiring on behalf of evangelicals. I ask that the book be taken seriously inside and outside the evangelical camp. It deserves to be, and in particular Donald Bruce and Don Horrocks are to be thanked for all the work that they have put into it.

Neil Summerton
Director, Oxford Centre for the Environment, Ethics and Society
Mansfield College Oxford
3 October 2001

Introduction

Genetically modified crops and food represent one of the most important and complex issues posed by modern technology. It is where genetic engineering 'comes down to earth', because, unlike many more specialised genetic applications, food affects us all. After some years in which relatively few concerns were expressed, GM suddenly burst into prominence in the UK in early 1999, following a media campaign by an alliance of environmental and consumer groups, development agencies and several newspapers. This led to a widespread supermarket and consumer rejection of successive governments' policies to introduce GM foods and promote the use of genetic engineering in crop production. This has caused ripples across Europe, North America and around the world.

These concerns were not always the case. The first main product in the UK was Zeneca's genetically modified tomato paste, introduced in 1996. It was clearly labelled and sold fairly well for several years. At this stage the most prominent UK environmental NGOs (non-governmental organisations) had not yet become seriously involved.

Genetic Engineering: A Symbol of the Times

According to consumer groups, a key change occurred when the European Union (EU) allowed the importing of unsegregated and unlabelled GM soya and maize from the USA.[1] This left consumers no choice but to eat food derived from genetically modified crops on a wide scale. To this was added the post-BSE factor, which has, not unnaturally, left a climate of scepticism towards foods which might be seen as at all risky or uncertain through human intervention. Since GM foods offered consumers no choice, potential risks and no tangible benefits, the rejection was predictable.[2]

Before the counter-claims and rhetoric began, various Christian groups were already examining the issues. In 1993 the Church of Scotland began a five-year expert study into the ethics of genetic engineering in crops, food, animals and micro-organisms, which led to the book *Engineering Genesis* published in 1998.[3] This foresaw a GM food crisis arising from a collective failure of the scientific and industrial communities, the UK government and the EU to take sufficient account of public values. This view was confirmed by other studies.[4, 5] GM food has struck a chord in society because it is an indicator for a set of much deeper questions.[6] These include human intervention in God's creation, the role of science, the right way to do agriculture, risk and precaution, global poverty and world food supply, as well as the power structures and inequities of the commercial and political world, globalisation, and public accountability.

---

1 House of Commons Research Paper 99/38, *Genetically Modified Crops*, Science and Education Section, House of Commons Library, 31 March 1999.

2 D. M. Bruce, J. T. Eldridge and E. J. Tait, "Genetic Risk Regulation, Society and Ethics", in *Risk in a Modern Society: Lessons from Europe*, Annual Meeting of the Society for Risk Analysis - Europe 2-5 June 1996, proceedings, p. 109, Centre for Environmental Strategy, University of Surrey, Guildford, 1996.

3 D. Bruce, and A. Bruce (eds.), *Engineering Genesis*, London: Earthscan Publications, 1998.

4 ESRC Global Environmental Change Programme (1999), *The Politics of GM Food, Risk, Science and Public Trust*, Special Briefing No. 5, University of Sussex.

5 House of Lords, *Science and Society*, Report of the House of Lords Select Committee on Science and Technology, HMSO, London, 2000.

6 D. M. Bruce and J. T. Eldridge, "The Role of Values in the GM Debate", in *Foresight and Precaution*, M. P. Cottam, D. W. Harvey, R. P. Pape and J. E. Tait (eds.), pp. 855–862, Balkema, Rotterdam, 2000.

Christian Involvement in GM Food Issues

At root, these are theological questions, and it is important that Christians consider them and provide a voice in the market place. Some churches had foreseen the need for ethical assessment in this area. Following the Engineering Genesis study, the Church of Scotland General Assembly adopted a policy in 1999 cautiously in favour of GM food and crops, but critical of abuses of power, control and regulation, and the failure of public accountability.[7] The Methodist Church has had an active interest for some years, and produced a discussion pack on genetic issues.[8] Other mainstream UK churches have become involved more recently, although thus far few have committed themselves to a specific policy. A Church of England Board of Social Responsibility briefing note advised in favour of GM foods,[9] but the Church Commissioners' advisory panel took a more precautionary position over the use of GM crops on land owned by the church.[10] The United Reformed Church has published a series of contrasting articles in its journal *Reform* over the period November 2000 to February 2001 to promote discussion.[11] Other groups operating across the churches have expressed particular viewpoints. Christian Aid are members of the alliance of groups

---

7 Church of Scotland, Reports to the General Assembly and Deliverances of the General Assembly 1999, *The Society, Religion and Technology Project report on Genetically Modified Food*, pp. 20/93-20/103, and Board of National Mission Deliverances 42-45, p. 20/4.

8 *Making Our Genes Fit – Christian Perspectives on the New Genetics, A Study Guide* (Peterborough: Methodist Publishing House, 1999).

9 Church of England Board of Social Responsibility, *Genetically Modified Organisms – A Briefing Paper,* London: Church of England, 1999.

10 Church of England Ethical Investment Advisory Group, *Genetically Modified Organisms,* London: Church of England, 2000.

11 D. M. Bruce, *Genetic Engineering – the Missing Values,* Reform, Nov. 2000, pp.16-17; T. Cooper, *GM Food Stop (…or Go)?, ibid.,* p.20, December 2000; J. Biggs, *GM Food (Stop) …or Go?, ibid.,* p.20, December 2000; K. Bundell, *Feed the World?, ibid.,* p.12, January 2001; D. Burke, A Window of Opportunity, *ibid.,* p.13, January 2001; N. Messer, *GM in the Balance, ibid.,* pp.24-25, February 2001.

campaigning against GM food, and in May 1999 produced a controversial report denouncing the technology from an overseas development perspective.[12] The environmental organisation Christian Ecology Link has also argued against GM on ecological grounds.[13] Amongst evangelical groups the John Ray Initiative has produced briefing papers expressing cautious support for GM food.[14, 15] Christians in Science have had conferences on genetic themes and the Agricultural Christian Fellowship have also studied GM issues.

The Evangelical Alliance has not hitherto addressed these issues directly. In 1994, however, it gave its support to the *Evangelical Declaration on the Care of Creation,* which is reproduced as Appendix 2, and as a result produced its own Creation Care leaflet.[16] This affirmed that Christians are called in the Bible to look after the earth with wisdom, compassion and justice. The earth is not ours; it belongs to God. Human beings are thus 'caretakers' or 'stewards' of God's earth. Whilst in Genesis 1.28 the Bible asserts human dominion over the earth, which has been misused to justify an attitude of uncontrolled exploitation of the earth, this is not the full biblical picture. The overall teaching of the Bible reinforces human responsibility to care for the earth, with wise regard for the impact of what we do in relation to animals, vegetation, health and environment, and our fellow men and women.

Against this wider biblical context, genetically modified crops and food pose important questions about our stewardship of God's creation and how we should use God's gifts in science, agriculture, commerce and government. How then

12 Christian Aid, *Selling Suicide.* London: Christian Aid, 1999.

13 Christian Ecology, *The Church of England's View on Genetically Modified Organisms – A Response by Christian Ecology Link*, Harrogate: Christian Ecology Link, 1999.

14 John Ray Initiative, *Genetically Modified Crops*, Briefing Paper No.5, John Ray Initiative, Cheltenham, 2001.

15 John Ray Initiative, *Christians and Genetic Manipulation – Are we 'Playing God'?*, Briefing Paper No. 6, John Ray Initiative, Cheltenham, 2001.

16 Evangelical Alliance, *Creation Care*, London: Evangelical Alliance, 1994.

should evangelicals in general be responding? Historically evangelicals have tended to welcome new technologies, but GM food brings together two sensitive issues – food and genetics. Despite the form of genetic engineering we have been doing for centuries through selective breeding in both plant and animal kingdoms, some today associate genetics with 'tampering with God's creation', and are dubious about atheistic brave new world scenarios. In a post-BSE context, anything scientific which threatens our sense of security about food is apt to be treated with great caution. GM raises some fundamental issues and it is already clear that Christians are divided over them. Some stress the opportunities of GM technology, others stress its threats, but almost all express concern over the power structures which have determined and controlled its application. This study by a working group of the Evangelical Alliance Policy Commission sets out some of the main issues in the light of Scripture, and of Christian insights into science, the environment, agriculture and the needs of developing countries.

About This Study

This study focuses on genetic engineering of food and crops. We did not seek to address questions of human or animal genetic engineering which raise many additional issues beyond our present scope. We also need to make clear what we mean by GM crops and food. In the past two years 'GM food' has become a label used rather carelessly to describe a range of actually quite different things. When we use the term 'GM crops', we refer to crops grown in the field or in a glasshouse in which genes have been added, transformed or disabled in a laboratory using the methods of molecular genetics, rather than selective breeding. As we shall see in Chapter 2, however, the difference between the two methods is actually more of a continuum than a strict boundary.

There is much debate about what the term 'GM food' ought to mean. Many scientists assert that it only means food that contains ingredients which are genetically modified, in which the novel gene construct or its product is actually present in what we eat. They draw a distinction from crop genetic modification which does not affect the content of the food. The more popular use is that 'GM food' refers to any food for which a process of genetic modification has been used at some point – in the preparation of the seed, the cultivation of the crop, or the food processing. Since some people interpret what is GM by association as well as by strict content, we have adopted the popular usage. We use 'GM food' to mean food deliberately produced involving a genetic modification at some stage, whether or not the food actually eaten has detectable quantities of the relevant gene or protein still present. We do not use it to mean those crops or foodstuffs which have incorporated small amounts of GM-derived material by accident, for example by unintended cross pollination in the field, or by becoming mixed in a silo with residues from a previous GM batch of the product.

Various evangelicals known to be involved in examining issues of genetically modified crops and food were contacted in autumn 1999. A working group was formed from those able to commit the time to a study. It was important to seek a balance of disciplines and perspectives, including theological, ethical, scientific, ecological, Third World and agricultural specialists, and to include both those broadly in favour and those sceptical of GM foods. A vital aspect was the submission of work in progress to critical peer review – technically, theologically and ideologically – again from several sides of the debate. We are grateful to our six peer reviewers for their input at several stages, both in person and by correspondence. We also received useful input from members of the Evangelical Alliance Council. The study began in early 2000. It was presented in draft format to the EA Council in September 2000, and was adopted by the Evangelical Alliance Policy Commission in February 2001.

Given the diversity of viewpoints concerning GM food, the intention of this book is primarily to inform and offer insights for evangelical Christians and others to evaluate these complex and controversial issues. Conclusions are drawn where we were able to form a consensus view, but we also felt it important to explore where we differed.

The chapters are each posed as a response to a question. *What is genetic modification?* (Chapter 2) summarises the scientific aspects of genetic engineering and plant breeding. *Is it wrong to do genetic modification?* (Chapter 3) reflects on theological insights related to genetic modification in agriculture. We then address more specific questions. *Is genetic modification too risky?* (Chapter 4); *Does UK agriculture need GM crops?* (Chapter 5); *Should we go Organic, GM, or what?* examines the debate between GM, organic and other forms of agriculture (Chapter 6); *Is GM just a Powerful Tool in the Hands of the Powerful?* (Chapter 7) looks at the role of commercial companies in the development of GM crops, linking the UK situation with the wider global issues posed in Chapter 8, *Do we need GM crops to help feed the world?* We then make some general reflections in Chapter 9.

The Policy Commission has subsequently formulated a set of resolutions and recommendations reflecting the Alliance's standpoint, based on the working group's report. These are set out at the end in Chapter 10.

Finally, a set of appendices provides a Glossary of special terms (Appendix 1); An Evangelical Declaration on the Care of Creation (Appendix 2); a summary of the UK Regulatory system for GM crops (Appendix 3); a more detailed examination of Old Testament law regarding mixing seeds (Appendix 4); and a list of Working Group members and Peer Reviewers, and their areas of expertise (Appendix 5).

Before embarking on this, however, we offer below some reflections concerning more general issues which have arisen during the study. These cover the role of science, different forms of rationality, and how far 'GM issues' are in fact peculiar to genetic modification or to what extent they existed already.

In the public arena, there is a great variety of responses to the debate. We list the main issues raised by genetically modified food and identify value criteria by which different groups have constructed the GM issue. The diversity of voices leads us to reflect on the problems of truth-telling and campaigning, the media and public understanding.

General Issues Relating to Genetically Modified Food and Crops

Science in a Post-modern World

GM food and crops poses some basic questions about the place of science in society in the third millennium. The climate of optimism after the Second World War has given way to a more ambiguous view, sometimes welcoming innovations, but often sceptical. This more critical mood has been prompted in part by some of the things that have gone wrong, but also by the fact that developments in the life sciences, such as genetic engineering and mammalian cloning, are reaching levels of intervention which challenge deeply held norms and assumptions about life.

Ironically, in a post-modern context, both the claims of Christ and the assumptions of science are challenged, since, in different ways, both represent truth claims about the world. Some Christians are indeed wary of science for this reason, seeing it as a challenge to their faith. But several of the authors of this book are among the thousands of committed Christians who are also active scientists, and who see no inherent contradiction in this. For them, science is part of God's calling to discover how the universe works and to ascribe praise to its Lord and Creator. Thus, Christians may see genes and the intricacy of their workings as wonderful examples of the psalmist's affirmation that we are 'fearfully and wonderfully made'.[17] Yet at

---

17 Psalm 139:14.

the same time they recognise the proper limitations of science to describing the material realm of creation. The spiritual dimension of life lies beyond its scope to prove or disprove. When a relatively small number of scientists attempt to claim otherwise, they have stepped outside science at that point and into their own personal beliefs. This represents 'scientism', which seeks to make science the supreme authority by creating an idol out of it, instead of understanding scientific explanations as just one way amongst many to describe natural phenomena. Science cannot 'explain away' God any more than knowing how a car works means that it cannot ever have had a designer.

Christians should, however, beware of two secular pressures arising from science. One is the drive of scientific determinism which, without Christian values to critique and guide what is done in the name of science, promotes scientific rationality as an idol. In a fallen world, science's ability to reduce things to their smallest parts can tend towards a mindset that sees things *only* at the genetic or molecular level, in terms of functions and scientific tests. Christians rightly oppose a technicist attitude in the world of genetics which loses sight of the wider dimensions, and which views living systems only in terms of their functions, and God's creation as a mere resource. This breeds a false optimism, where exaggerated claims may be made about the latest technical breakthroughs, through hubris, naivety, or a desire to attract funding, while being concurrently short-sighted about drawbacks and unintended effects.

Less familiar is the opposite concern reflected in a rise in neo-pagan views which regard the scientific endeavour as deeply flawed and exaggerate the concept of pristine nature. Again, contrary to a Christian understanding of creation, nature is seen in divine terms, or at least not to be tampered with, fearful that 'she' will 'hit back at us'. The idea that 'nature knows best' tends to oppose human intervention as risky, or worse, not on the basis of human moral failure, but on false notions of 'harmony' derived from idealised, pantheistic, and ultimately idolatrous views of 'Nature'. This is examined further in the theology chapter. The Evangelical Declaration on the Care of Creation makes clear that Christians should be the first people to stress care for the environment, but a balanced Christian outlook also understands responsible human scientific activity in creation as valid, despite being tainted by the Fall, in the light of Christ's redemption and resurrection, when guided by the wisdom of the Holy Spirit.

The Church of Scotland study previously mentioned pointed out the exaggerated claims of objectivity on behalf of genetics, and the arrogance of some scientists, industry and government representatives in dismissing anyone with value-based objections as 'irrational'.[18] No scientist and no science is value-free. In practice, everyone approaches these issues with their own presuppositions and 'baggage'. What is important is transparency concerning what these are, whether opponents or proponents. Scientists have their own inherent values which underpin and affirm what they are doing, but are often unaware of the fact, because there is a tendency to regard any personal beliefs as inappropriate to their practice of science. On the other hand, some assume that all the scientific experts who espouse GM do so because they have financial interests in promoting the research. It would be far from the truth to suggest that most of those who promote GM crop research in the UK are motivated by greed and are financially obligated to the biotechnology companies. The primary motivation for most serious scientists remains curiosity and discovery. Others assert that the curiosity factor is seen too naively, or that experts have such a strong commitment to their position that it biases their opinion. Although this may be overstating the case, this risk is a major reason why we believe scientists need to be trained in wider ethical issues.

Rationalities

Because science is not the only way to see the world, we welcome the insights of other disciplines, notably the social

18 Bruce and Bruce (1998), *Engineering Genesis*, op. cit.

sciences, and the views of ordinary people. The attitude that presumes that scientific modes of thinking must be supreme on GM issues and all objections are irrational represents an abuse of power.[19] This mindset ignores Weber's important distinction between scientific rationality and ways of reasoning based on values.[20] Work based on exploring the views of non-experts through focus groups and other means has confirmed that ordinary people may have a well grounded basis for their concerns, rooted in basic values, even if they are not necessarily in possession of the 'facts'.[21]

The so-called 'yuk factor' in response to certain startling innovations, like inserting a fish gene into a strawberry, is often rejected as an emotional reaction based on unfamiliarity. Heart transplants are often cited as an example where people got over their initial repulsion. Yet on examination, such reactions may reveal that deeper value-questions are being challenged, and which are not 'mere emotion'. It is well established that the perception of risk by lay people may also differ greatly from the way a scientific risk assessor may evaluate it. We discuss this further in Chapter 4. This underlines the importance of opening the processes of decision making and debate about biotechnology in general to a much greater degree of public participation. We discuss this further in Chapter 8.

On the other hand, there are limits. At the height of the GM crisis, many people were repulsed at the idea that with GM food we might be 'eating genes'. They did not appreciate that all of us eat genes every time we eat a raw fruit or vegetable.

---

19 ESRC Global Environmental Change Programme (1999), *The Politics of GM Food, Risk, Science and Public Trust*, Special Briefing No. 5, University of Sussex.
20 M. Weber (1918), 'Science as a Vocation' in: H. Gerth, and C. W. Mills, *From Max Weber*, Routledge and Kegan Paul: London, 1948.
21 C. Deane-Drummond, R. Grove-White and B. Szerszynski, Genetically Modified Theology: The Religious Dimensions of Public Concerns about Agricultural Biotechnology , in *The Reordering of Nature: Theology and the New Genetics*, edited by Celia Deane-Drummond and Bronislaw Szerszynski (with Robin Grove-White), Edinburgh: T. & T. Clark, 2002.

This also raises an important point concerning what is truly 'new' about genetically modified food.

Are GM issues really 'nothing new'?

Food production has become separated from most consumers' experience. At the start of the twentieth century most British people still had family links to farming and the land; by its end, few have anything to do with food until they pick it off a supermarket shelf. Consequently, many agricultural and bio-technological innovations over the past century have escaped the notice of most city dwellers. GM food brought to light many questions that already existed but were not generally recognised. For example, oil seed rape had already been bred with genes to resist herbicides by selective breeding, and was subject to far less rigorous testing than occurs with GM crops today. Christians need to be clear that some issues, like gene flow or biodiversity loss, have not suddenly been thrust upon us by GM, but were present long before the advent of molecular genetic modification. Indeed, often GM serves as a prompt to go back and look critically at existing practices in agriculture which were not subjected to public scrutiny at the time.

While it is important to distinguish what are the issues which GM raises uniquely, it would be a mistake to see it in isolation from its wider societal context, and especially the assumptions of the societies which bred it. As we shall see throughout this study, GM issues are deeply linked to the power structures of commercial sectors and government, the regulatory framework, the forces of market and cultural globalisation, the global mismatch of rich and poor, and much else. Many of the concerns which GM raises are indeed much wider questions. It is important to identify which particular aspects GM is highlighting, and to consider what impact these have on existing problems and injustices, or the measures to address them. GM is a package of issues which need to be taken together.

Different voices and different criteria, different ways of constructing the GM issue

GM food is therefore not a single question about which one is 'for' or 'against'. It encompasses a very complex set of ethical and social issues, which we summarise in a series of questions at the end of this chapter. It is important to recognise the post-modern context of the GM debate with its diversity of voices, like vendors calling their wares in a market place of ideas. Without an overarching concept of truth, each is apt to claim that their way of framing the issues is the crucial one. This leads to widely differing conclusions. It is important to discern the value assumptions and criteria which lie behind each of these viewpoints. These include:

- Inherent ethical/theological/philosophical – is switching genes intrinsically wrong, or right?
- Ideological – e.g. technological or natural; conventional or organic agriculture?
- Risk – can we handle the genetic engineering of food with enough safety and environmental care?
- Scientific rationality – technological progress for human and ecological benefit?
- Commercial – how much economic growth, jobs, profit, competition?
- Farming – what does the UK farmer most need at this point in time?
- Resources – will GM feed an expanding world population with diminishing soil and water?
- Development – does GM equate to global justice for the poor, or the opposite?
- Control, power and participation – amongst corporate power, nation states, EU, World Trade Organisation (WTO), NGOs, activists and consumers, who should decide, and against what criteria?

As Christians, we recognise important points in each of these perspectives and criteria. Against this fragmented picture, Christians believe that Jesus Christ is the one in whom all things hold together. Christians consider that the biblical testimony should be able to provide a framework in which to make sense of these claims and counter-claims. Can we bring them together into an integrated and balanced whole? Which of these are particularly important, and which less so? That is a tall order indeed, but one which we must at least begin to address.

Media debate

We express our concern at the way in which the debate has often been carried out. The polarisation and politicisation of the issues during the past 18 months have not favoured constructive debate. The confrontational style of the news media and the campaigning approaches of different interest groups have resulted in hype, exaggeration and selectivity with the truth. We offer two examples.

Allergy trials are conducted on most potential novel foods, both GM and conventional. At a fairly early stage in development, soya beans carrying a gene from the Brazil nut were found to contain an allergen, and the company concerned stopped all further development of this gene. Consider the potential media stories. It could be presented as a case illustrating the potential risks of GM food, or of the care of the company concerned to test its products, or of the effectiveness of the standard regulatory system. In the wake of BSE, food scares are a soft target in media terms. (At the height of the 1999 crisis, two leading UK newspapers conducted a circulation war over which one was more 'anti-GM'. In such a context it was sadly all too easy to present this merely as an example of how dangerous GM foods are.)

As a second example, another newspaper carried a headline claiming the presence of human genes in food, and referring to the sheep at the Roslin Institute which are genetically modified with a human gene to produce a potential therapeutic protein in their milk. The article made no mention of the fact that in 1993 the Government had set up a committee to look into the ethics of this very case, which had led to the banning of the use of any such experimental GM animals in food, even those which did not contain the added gene.[22] The story was therefore simply untrue.

Not all reporting has been as misleading as this, but it underlines the severe limitations of the media to convene a serious public debate on issues as complex and sensitive as those to do with GM food. At their best, the media can alert the public to new discoveries and new problems alike, and draw out the issues. At their worst, competition may dictate that being first on a scoop matters more than that the story was largely false. What is perceived as 'good TV' may mean taking only the most extreme views, even if they represent only a handful of people.

Truth-telling and campaigning

This raises a wider question for Christians about truth-telling and campaigning. Exaggeration for the sake of effect is deeply ingrained into the campaigning mentality. This occurs on both sides – in 'pro-GM' claims made by some companies, scientists and Government spokespeople, and 'anti-GM' assertions on the part of some NGOs. For the former there may be the share price, the funding, or the party line to consider, or they may simply be carried away with their enthusiasm. For the latter there is the desire to make the point passionately believed in, and a pride in being there first, for having been first to react radically, for getting the issue on the public agenda, even though the case might be misrepresented or relatively unresearched. For Christian organisations and individuals, this poses problems of truth. Being economical with the truth flies in the face of New Testament teaching. Paul states, 'We have

22 Ministry of Agriculture, Fisheries and Food, *Report of the Committee on the Ethics of Genetic Modification and Food Use*, (Polkinghorne Committee report), HMSO: London, 1993.

renounced secret and shameful ways; we do not use decep-
tion… On the contrary, by setting forth the truth plainly we
commend ourselves to every man's conscience in the sight of
God.'[23] For the sake of a campaign and the best impact, there is
often a tendency to choose only the data and examples which
support the campaign goals, and not to present the full range of
evidence. Experience suggests that what most ordinary church
members seek is simply a balanced portrayal of the issues so
that they can make up their own minds. Christians of all people
should beware most of the post-modern tendency to represent
facts and truth as less important than 'the agenda', 'rights', or
'political correctness'.

Public responses can often be made on the basis of intu-
ition and impression, rather than a thorough appreciation of
the questions. Out of the confusion, a serious issue is emerging of the need to find a trustworthy source amid all the dis-
parate voices. Christians could have a vitally important role to
play in society at this point. One of the main aims of this
study is to stand back from any particular partisan stance, to
bring clarity to the various issues, and to offer perspectives
which go beneath immediate impressions. We also seek to
determine how far attitudes and impressions arise from limited knowledge, poor exegesis or secular influences, and how
far they are the product of a rigorous and informed biblical
faith.

Summary of the issues

In the time and space available an educative document which
covered all the above ground would be a huge task. Inevitably
this book represents something of a snapshot, partly because
the associated technology and its social and political aspects
change by the month. The basic issues, however, are fairly
clear-cut, and we can summarise these as a series of important
questions:

---

23 2 Cor. 4:2.

- Does genetic engineering represent an intrinsically right or wrong intervention in creation? Against what criteria should we decide – what is our theology of creation, human intervention and science; do any Scriptures relate to the specific case? (Chapter 3)
- Do we have the skills, insight and foresight to manipulate the fundamental nature of creation? In what sense, if any, is this different from any other human technological intervention? Does our fallen nature make any special difference in the case of GM food? (Chapters 2, 3 and 4)
- Do GM crops pose risks to health or the environment that are significantly greater than those posed by conventional selective breeding or organic methods of agriculture? (Chapter 4)
- For what purposes, if any, should GM crops be used? (Chapters 5, 6, 7 and 8)
- How can we advise and support Christians who are involved professionally in farming and biotechnology? What do genetically modified crops mean for the farmer in the UK? (Chapter 5)
- Does genetic engineering represent a reasonable or misguided approach to agriculture, in the context of the current debate about organic and conventional approaches? (Chapter 6)
- Could GM crops play an important role in feeding the poor of the world, or would it be to their serious disadvantage? (Chapter 8)
- How do we interpret the issues of power and control in biotechnology? How can it be made more publicly accountable? (Chapters 7 and 8)
- What can Christians do to promote a constructive debate over GM issues in society? (Chapter 9).

2

What is Genetic Modification?

The Science of Plant Breeding and Genetic Modification

The technical terms shown in bold type are defined in the Glossary.

Introduction: Genes and Genetics

The term **gene** means an individual unit of the genetic material that specifies an individual characteristic in a living organism. For many genes it is possible to describe that characteristic in straightforward biochemical terms, for example the ability to make growth hormones. Genes are made of a chemical called **DNA** which consists of four types of building blocks, many thousands or even millions of which are joined together to make the long string-like molecules of DNA. It is in the order of these four chemical building blocks of DNA that genetic 'instructions' are written (just as it is in the order of the 26 letters of our alphabet that whole books are written). An individual gene is thus also a specific piece of DNA which is itself part of a much longer DNA molecule, a string of genes, that makes up a **chromosome** (Fig. 1). One of the most remarkable features of this 'language' of the genes is that it is universal, implying a common genetic heritage of all living things. Thus a simple bacterial cell can read the language of a gene from a plant or from a mammal. Indeed, it is this central

Figure 1. Active plant genes viewed with the electron microscope. The active genes are separated by stretches of the DNA 'string' that do not function as genes; these are the spacer DNA between genes. The active genes have a 'bottle-brush' appearance because they are being actively copied to give many 'messengers' that will actually convey the genes' instructions to the cell. At the 'pointed' ends of the bottle brushes copying has just started: the messenger molecules are still very short. At the 'blunt' ends, copying is nearly finished and the messengers will shortly 'peel off' from the genes. The scale bar represents 0.5 micrometres (a micrometre is 1/1000 of a millimetre).

fact which makes many of the applications of genetic modification possible.

Although all members of a particular species have the same
set of genes, many of the genes actually vary slightly in structure between individuals. In the vast majority of cases, this
makes no difference to the animal or plant at all. The slight differences in gene structure have no effect on gene function.
However, in some instances, the genetic differences actually
make a difference that may be visibly obvious or that may be
detectable by biochemical tests. As humans we belong to a
variable species. To scientists this variation is a cause of wonder,
and to Christians we are reminded that the Creator cares for
each one of these members alike in all their differences and diversity, all over the world.

Amongst plants, the cabbage, *Brassica oleracea*, is also very
variable, so much so that the different varieties have different
common names (such as sprouts, broccoli and cabbage). However, many species are not nearly so variable as humans or cab-
bages and it is such situations that we will consider in relation
to plant breeding. Before doing that, we need to introduce two
more terms.

We know from studies of the way that genes work that genetic changes occur because of changes, often very subtle, in
the arrangement of the building blocks of DNA to which reference has already been made. These changes are known as
mutations (from a Latin word meaning to change). Sweet
corn, for example, is a mutant of maize that is very inefficient at
turning sugar into starch. The new super-sweet varieties are
double mutants. Some mutations may lead to a loss or impair-
ment of gene function that is actually harmful or at the extreme, lethal to the organism. The apparently benign mutation
in sweet corn leads to the seeds being less efficient at germinat-
ing, because there is less starch to feed the growing seedling.
Many mutations, however, may be harmless or even beneficial
to the organism and often become established in wild populations. This leads to the genetic variety that we have mentioned
already. Indeed, *all* genetic variation comes from mutation of

the organisms' 'original' genes. Where more than one version of a gene becomes established, the different forms are known by geneticists as **alleles**. The relative abundance of different alleles of a gene in a population may change if one allele is for some reason favoured over another. This process, of which there are many known examples, is known as natural selection, and this is one of the basic 'tools' of the plant breeder.

Historical Aspects of Plant Breeding

It is clear from archaeological records that humans have been collecting, saving and planting seeds for over 12,000 years. It is also clear that, consciously or unconsciously, these early farmers made use of the genetic variety that we have mentioned above. Thus they would save seeds from plants that had desirable characteristics (e.g., from grasses that did not shed seed, from higher yielding plants, from those that survived drought, disease, etc., and produced a harvestable yield). This was the start of artificial selection and plant breeding, and its effects on certain crop species was quite dramatic. Archaeological surveys of the Anatolian plateau show for example that selection for yield led to significant increases in the number and size of grains per ear in wheat, all this happening very rapidly without any deliberate cross-breeding between what we would now call 'varieties'. Humans were in effect, without knowing it, making use of natural selection.

This type of selection and the planting of self-saved seed proceeded for many centuries, and continues today amongst some subsistence farmers. The variation in alleles and new combinations of them have led to the emergence of distinct local forms of cultivated crops, known as landraces, many of which have survived to the present day. Some of these landraces are not very high yielding, but they are often well adapted to the particular environment in which they are traditionally grown (e.g., drought tolerance, disease resistance, or low nutrient requirements). These characteristics are all conferred by the combination of alleles that are in those plants, with the crop as a whole in that area having a high genetic diversity. This genetic diversity is an important local resource, and potentially also to plant breeders in other countries. The accessing of these resources by transnational corporations has, however, become a deep source of contention. Local people have often felt themselves to be unfairly exploited. It is widely seen as important that landraces are conserved both as a resource for subsistence farmers and as a general resource for plant breeders. This is done in seed banks in a world-wide network of international crop research stations, intended for the benefit of all, but there are fears that some of these may pass into private ownership through lack of public funding.

Wherever plants are gathered together and sown as crops, there are immediate problems. Weeds compete with, and pests and diseases live off, crop plants. This lowers the yield obtained from any given area of crop. Pests, such as locusts, and epidemics of plant diseases, can both wipe out a crop over a wide area. In UK intensive farming this results in reduced profits, and possibly bankruptcy. Elsewhere in subsistence farming, it can result in starvation. The farmer has always tried to control these pests and diseases, although understanding the origin of plant disease was very rudimentary. During the nineteenth century in Europe it was confirmed that a variety of micro-organisms caused disease. This changed the approach to pest and disease control whilst discoveries in genetics did the same for plant breeding.

The Science of Plant Breeding

The nineteenth-century monk Mendel showed how plant characteristics could be inherited as self-contained characters. His work pre-dated Darwin but was lost for some years, and Darwin was unaware of it. It was rediscovered early in the twentieth century and applied very quickly to plant breeding. The yield of wheat varieties was improved and the straw was shortened considerably. As the varieties that came onto the market improved, the search was widened for more desirable characters to insert into particular crops. This search moved from landraces to closely related wild species of plant. This type of plant breeding made use of existing genes in the crop and in wild relatives of the crop.

Thus, the breeder is constantly looking for new genetic variation in a particular crop species in order to generate new varieties that are better in some way. The 'classical' breeding techniques that we have described did not necessarily give the desired variation and characteristics, however. In some cases, **hybridisation** of the crop species with a wild species is attempted. Plant breeders often accelerate the rate at which variants occur by deliberately inducing mutations using radiation or special chemicals. This is a very drastic process and kills a high percentage of the seeds that are treated, but the survivors very often show useful genetic characters. Thus, the variety of barley used to make nearly all the beer in the UK is a mutant generated by radiation, as are several of the high-yielding rice varieties.

Many of the hybridisation methods that are used actually require significant human intervention to achieve success. Plant breeders employ a variety of techniques in order to by-pass fertilisation barriers between species, and so facilitate some of the cross-species hybridisations mentioned above. A spectacular example of this is that plant breeders have managed to incorporate a segment of a rye chromosome containing many genes into one of the wheat chromosomes and a gene conferring resistance to eye spot from a wild grass species, *Aegilops ventricosa,* into another. These transfers were not achieved by the processes that we call 'genetic modification', but nevertheless would certainly not occur in nature. This breeding work with wheat led to enormous increases in yields in the 1970s. The fact that people in the UK today can eat wholemeal bread made with home-grown wheat is thus the culmination of applied research in plant breeding, making use of a range of novel 'interventionist' techniques. Indeed, one of the aims of introducing a segment of rye chromosome into wheat in the 1960s was to 'import' disease-resistant genes from rye to wheat. This illustrates that the potential for plant breeders to improve a crop is limited only by the gene pool of the crop, although the gene pool of undomesticated related species may be useful in some cases.

At their most extreme, these 'non-GM' breeding techniques include the fusion of cells or even **nuclei** (the parts of the cell containing the genetic material) of different species, a process that has been used to generate new varieties of ornamental plants.

Between 1920 and 1960 many advances were made in yield and quality, but it was much harder to introduce pest and disease resistance into crops, and it proved impossible to control weeds via plant breeding.

Furthermore, conventional plant breeding is an imprecise process. Several thousands of crosses have to be made to obtain one or two useful 'new' varieties with the right characteristics. Selection has to be carried out over many generations. Candidate varieties in the UK have to be submitted for trials in order to be accepted onto a 'national recommended list' before they can be marketed. Before a new variety can be marketed it has to satisfy an independent panel (National Institute of Agricultural Botany, NIAB) that it 'breeds true', that it has all the characteristics that it claims, and that it performs well over a wide geographical range. It also has to be registered for **plant breeders' rights**, which provide royalties on the variety but do not restrict the right of other breeders to use it in further breeding experiments.

With conventional plant breeding, when crosses are made an unknown number of genes, typically 1,000 to 2,000, of largely unknown function will be transferred into the candidate variety along with the wanted characteristic (gene), even when the hybridisation is between two varieties of the same species. Some of these unwanted genes may have deleterious side effects in the plant, like reduced yield or susceptibility to disease, or may possibly affect humans. Some otherwise excellent varieties of potato contain excessive amounts of natural insecticides known as glyco-alkaloids which are potentially poisonous to humans. A new variety of celery nearly reached the market place before it was discovered that one of the unwanted characters that entered during the plant breeding programme caused allergic reactions in many people. Once the desired character is established in the candidate variety, further selection takes place seeking to eliminate these unwanted genes. If the unwanted characteristics cannot be eliminated, the variety is excluded from further development. All this means that new varieties take at least ten years to reach the market. However, the plant breeders' rights only last for 25–30 years after entry into the market place. Hence, all costs of the research and breeding programme as well as profit to the breeding company have to be recouped in that time.

Genetic Modification

Potentially, genetic modification of crops offers a way round some of the problems faced by the plant breeder. Before discussing these applications in plant breeding, it is necessary to look at the basics of the process. Genetic modification is also known as 'genetic manipulation' and 'genetic engineering', and was invented in 1973. The technique was based on the *natural gene transfer* mechanisms that occur in bacteria and certain other micro-organisms. It depends on the fact that micro-organisms can transfer small pieces of DNA called plasmids between cells (Fig. 2). The techniques also use specific biochemical catalysts (enzymes) which can cut and re-join DNA molecules in a very precise way. This enables the scientist to make *recombinant DNA molecules*.

The technique was very rapidly taken up by the pharmaceutical industry. This is exemplified by the deliberate transfer into bacteria of a human gene that normally instructs pancreas cells in the human body to make insulin (Fig. 3). In 1982, insulin produced by genetically modified bacteria was licensed for

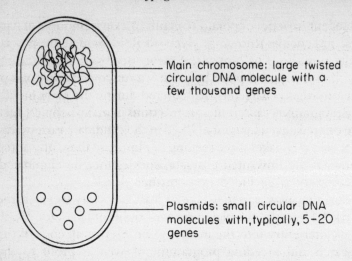

Figure 2. Many bacteria contain small 'extra' pieces of DNA that are usually circular (i.e. the DNA 'string' is joined into a circle). These are called plasmids. Note: the plasmid and main chromosome are not drawn to scale.

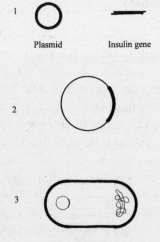

Figure 3. Diagrammatic representation of the cloning of the insulin gene.
1. The insulin gene has been identified and isolated and is now ready to be inserted into a plasmid (see Fig. 2);
2. The plasmid has had the insulin gene inserted into it;
3. The plasmid is transferred into a suitable bacterial cell (note that the bacterial cell retains its own main chromosome). This will enable the bacterium to make insulin. Not to scale.

use in human therapy. As indicated earlier, it is the universal genetic language that makes such changes possible and effective. The bacterium 'reads' the human gene as if it were one of its own and makes the product, i.e., a human hormone, which that new gene tells it to, and which the bacterium had never made before. The fact that the insulin gene came from human DNA does not make the bacterium human. It is still very much a bacterium. This success has been followed with the production of many therapeutic proteins and several vaccines in genetically modified bacteria and other micro-organisms.

The technique is used to produce an enzyme used in the manufacture of a sweetener for 'diet' fizzy drinks, and another enzyme called 'chymosin' for use in cheese making. Cheese made this way is sold as 'vegetarian cheese', because it avoided the need to extract the enzyme from calves' stomachs. It is strange that with all the concerns about GM foods and crops, these applications seem to have escaped attention, especially since the product of the genetic modification process is consumed with the cheese.

Research uses of genetic modification

The ability to transfer genes from any living organism into cells of micro-organisms opened up a research use for GM. This has grown enormously in scope and sophistication over the past 25 years. The current knowledge about gene structure and function could not have been dreamt of prior to the invention of GM. The Human Genome Project would not have been even conceivable without the ability to copy individual human genes and groups of genes via micro-organisms, in order to provide material for analysis. Work on plant genes has also benefited from the application of GM techniques, as is evidenced by the recent publication of the complete genome sequence of a small weed, *Arabidopsis thaliana* (thale cress), a plant used extensively as a model for all plants in studies of genetics. The rice genome has now also been published. One of the present authors also routinely uses GM techniques in his research on plant genes. There

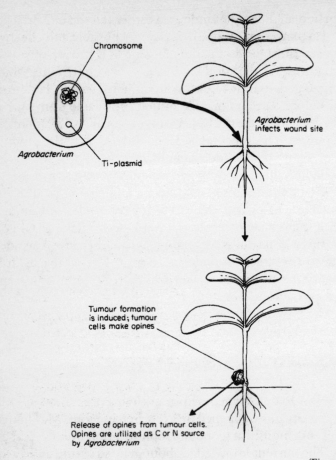

(Figure 4a)

Figure 4. Infection and transformation of plant cells by a soil bacterium, *Agrobacterium tumefaciens.*

(a) A generalised diagram of the infection process. Note that the bacterium causes 'unscheduled' division and growth of the host cells, thus causing a gall or tumour. The infection also results in the cells in the gall making bacterial nutrients called opines. The plant cannot use these and thus the plant cells have become 'factories' for making bacterial nutrients. Not to scale.

(b) Like many bacteria, *Agrobacterium* has a plasmid. However, this plasmid is much larger and contains more genes than we described in Fig. 2. The infection process involves transfer of part of the plasmid, the T-DNA, into the DNA of the host

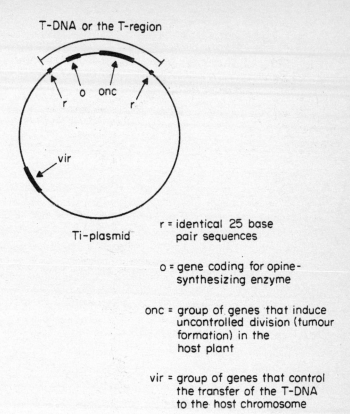

T-DNA or the T-region

Ti-plasmid

r = identical 25 base
    pair sequences

o = gene coding for opine-
    synthesizing enzyme

onc = group of genes that induce
      uncontrolled division (tumour
      formation) in the
      host plant

vir = group of genes that control
      the transfer of the T-DNA
      to the host chromosome

(Figure 4b)

Figure 4.(*continued*)   plant. This T-DNA is defined by two short 'borders' (r on the diagram) and contains the genes necessary to 'drive' the plant cells to divide and grow into a gall (onc on the diagram) and the genes necessary to enable the plant cells to make opines, as in Fig. 4a (o on the diagram). Note also that a group of genes elsewhere on the plasmid, the 'virulence' genes, are necessary for this DNA transfer process (vir on the diagram). Not to scale.

Scientists discovered that the 'harmful' genes (o and onc in 4b) could be deleted without preventing transfer of the DNA to the plant cell. The essential regions for transfer are the borders (r in the figure). Any DNA inserted between these borders will be transferred to the host plant in the now harmless infection process and new plants carrying the inserted gene(s) can be grown from the infected cells.

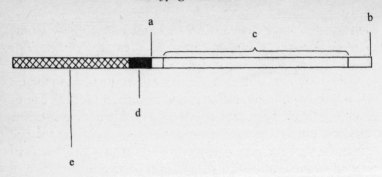

Figure 5.   The structure of a typical gene such as may be transferred to plant cells by genetic modification. The gene itself lies between a and b in the diagram. The gene is copied into messenger molecules (see Figure 1) in the a to b direction. The region marked c is the region of the gene that is actually 'de-coded' by the cell as a specific instruction. The gene control region, known as the promoter and lying 'upstream' of the gene, is marked by d and e. The essential switch mechanism (known by scientists as the basal promoter) without which the gene cannot be switched on, is marked by d; the region marked e is a specialised region of the promoter that defines where and when d is 'allowed' to switch the gene on. Thus a scientist might ensure that a gene inserted into a plant by GM techniques was only switched on in leaves or seeds or in response to drought (or whatever...).

is now a detailed knowledge of the position of genes in several crop species, especially in the cereals. This knowledge is being used to achieve gene transfer between cereal species, by both conventional plant breeding and GM techniques. This contribution of basic GM techniques, and the array of methodologies that grew from them, has been of huge value in basic and applied research in fields as diverse as medicine and agriculture. This serves as an important context against which to consider the genetic modification of plants.

Plant Genetic Modification

The initial development of plant genetic modification also relied on natural mechanisms, and especially on the manner in which a particular bacterium, *Agrobacterium,* transfers some of its genes to plants. **Agrobacterium** is a natural pathogen of plants that enters the plant through wounds. The bacterium then transfers a small piece of DNA, containing copies of some of its genes, to the host plant cells at the site of the wound. The transferred genes become integrated into the plant's DNA and provide the genetic information that causes the plant to form a gall and the cells in the gall to produce nutrients for the bacterium (Fig. 4a,b). This is effectively a form of genetic parasitism. Scientists quickly discovered that the genes that caused gall formation and the genes that caused production of bacterial nutrients were not needed. All that was needed in this piece of bacterial DNA were the short building-block sequences that allow the piece of bacterial DNA to insert itself into the plant chromosome. Between these two short sequences the scientist is now able to insert any gene, along with its on–off switch mechanism (known as the promoter), and then to let the bacterium transfer that to the target plant (Fig. 5). Indeed, in some of more 'minimal' versions of the mechanism, the bacterium has been dispensed with altogether and the DNA is 'shot' into the plant cells. Again, the universal genetic language allows the insertion of genes from almost any other living organism – but the plant is still a plant.

This technique is both precise and imprecise. It is precise because, unlike plant breeding techniques, only one or a few specific genes which confer the desired characters are transferred to a plant. The rest of its genetic characteristics are not altered. Single wanted genes can thus be moved with precision into candidate varieties. This clearly contrasts with what happens in conventional breeding, described earlier. However, genetic modification is imprecise because of position effects. There is no control over the place where the 'incoming' genes are inserted within the plant chromosome. This causes great variation from plant to plant in how much the desired character is actually seen in this first generation of the modification. It means that there must be some selection of the first generation of GM plants, followed by observation of the stability of inheritance in subsequent generations. However, this is shorter than the sorting and evaluation phase in conventional breeding, and leads to a faster adoption of the new varieties.

The technical advantages of GM are that it enables the addition of specific genes to well-characterised varieties, that those genes can come from widely separated species, and that new varieties may more quickly be adopted for use. The main advantage is that it can increase hugely the genetic variety available to the plant breeder whilst avoiding the problem of bringing in unwanted genes. Of immediate relevance to the grower is the ability to insert genes for resistance to pests and diseases. Thus there are now maize, cotton and oil-seed rape varieties that are resistant to particular insect pests, and there are a number of cash crops (e.g. tomato) in which new GM varieties are resistant to viruses. Moreover, the ability to handle individual genes in the 'test tube' means that direct biochemical modification may be used to generate new variants in existing genes within a crop species.

The term 'genetically modified' encompasses a wide range of types of change. In many actual and potential applications of plant GM, the individual genes that will be transferred in will not cross wide species barriers if there is no advantage in doing so. Indeed, some applications of GM do not involve mixing genes across species at all. Some involve transfer of individual genes between different varieties within the same crop species. Others increase the number of copies of a gene within an individual crop variety. It is also possible to 'switch genes off', or even to put back a gene after it has been biochemically manipulated in the test tube.

Although much has been made of inter-species gene transfer, the distinction between 'conventional' plant breeding and GM techniques can be somewhat blurred. For example, short-stalked wheat was first obtained by conventional breeding in the 1970s. Now the allele that causes the short stalk has been identified and isolated and it has been transferred to rice using GM techniques, without the problem of bringing in unwanted characters. Secondly, herbicide-tolerant oil-seed rape has been obtained both by gene mutation in conventional breeding programmes and by GM techniques.

Scientific Aspects Relevant to Some Objections to GM Crops

It is 'unnatural'

Although gene transfer mechanisms do occur in nature, and indeed have been made use of in the development of GM techniques, it is obvious that use of GM requires extensive intervention. It uses processes derived from nature, otherwise it would not work, but it is nevertheless, in one sense, 'unnatural'. The same processes, however, would be applied to many practices used in conventional breeding. But GM may lack many of the side-effects of conventional breeding, like unwanted genes. As we have pointed out above, the distinction between selective breeding and GM is clear at the extremes but blurred in the middle. It is therefore important, if arguments based on 'naturalness' are used, to be quite precise about what is objected to, as to whether it is truly 'unnatural'. There is a tendency, for example, to equate 'natural' to a memory of what things were like in one's childhood. As we observe in the next chapter, little that we rely on for normal life could be fairly described as 'natural' in a pristine sense of the word – the concept is ambiguous in Christian ethics. However, there are other potential grounds for intrinsic objection to gene transfer across species which we examine next.

Uncertainties in the genetic method

It is frequently suggested that we have rushed too fast into GM crops whilst not enough is known. For the scientist it is difficult to know how much experimentation is necessary or desirable to ensure that 'enough is known'. The timescale involved in the commercial production by GM techniques of human insulin was very short at six years, compared with Monsanto's herbicide-tolerant soya beans which took 11 years and which have now been grown in the USA for four years. The question is how much longer would it take for any novel effects from the gene construct itself (as opposed to environmental impacts) to manifest themselves?

A prominent scientific advocate against genetic modification is Dr Mae-Wan Ho and the Institute for Science in Society (ISIS). She maintains that genetic modification is inherently unsound as a technique. She objects to the reductionist assumptions of genetic engineering, and asserts that to take genes out of one context and insert them into another is inherently likely to cause unpredictable and disastrous effects, and that the vectors used to transfer genes are especially dangerous. Her objections are intrinsic, grounded in her personal beliefs of an alternative, holistic nature. Genes are seen as so much a part of the wider web of nature, interconnected with the whole of the rest of the universe, that moving them will disturb the distribution of the 'vital force'. This leads to a rejection of a reductionist view of genes, a position with which we have some sympathy. But in doing so there is no acknowledgement at all of the very well-founded paradigm that individual genes have individual biochemical functions that can be studied individually. Thus, while she is right in criticising a tendency to assume that genes can be moved around as though they were completely independent of their environment, the question is one of degree, as to how far and how often this would lead to serious unintended consequences. Many of the dire predictions she has made in her writings have drawn criticism for being exaggerated and speculative. If one does not share her intrinsic objection, however, it is prudent to keep an eye open for evidence of unexpected effects, but not to regard them as inevitable in the way that her theories assert.

What are the Applications of Plant Genetic Modification?

It is important to distinguish clearly between GM *crops* and GM *foods*. The 'vegetarian' cheese referred to earlier is clearly a food that is produced directly by a process that involves GM.

Further, as already noted, the component produced by GM, that is, the chymosin enzyme, remains to a large extent in the cheese and thus becomes part of the food.

In crop modification, modifications which directly affect the food component of the crop may come into the category of GM foods. There are others which affect plant characteristics that have nothing to do with the food component. The slow-softening tomato, which was the first GM crop to come to market, is an area where the GM process affects the quality of the consumed food product. There are other applications under development that also directly affect the food product. However, it is very likely that only a relatively small proportion of the modifications will affect directly the food component of the crop. Many applications will affect the ability of the crop to withstand environmental and biological stresses, while others will affect the way that a crop is grown and harvested.

A third category is in the production of foods which are claimed to have enhanced nutritional qualities or additives which are designed to confer health benefits, in which the boundary between food and medicine becomes blurred. Some of these are promoted as 'nutraceuticals' or 'functional foods'. Some of these have been made with conventional technology, but in others genetic modification is used. One example of nutritional enhancement which has aroused much interest is the so-called 'Golden Rice'. A variety of rice has been modified by adding various genes in order to enhance its vitamin A content. A related example produces increased iron levels in rice. Both of these are aimed at helping to address dietary deficiencies in developing countries, which are the cause of blindness and certain diseases, particularly among children. It is also significant because it is a GM application developed in the public sector, primarily with academic and charitable funding, and which is not aimed at western markets. Another example is the possibility of increased vitamin C content to provide increased levels of antioxidants for protection against cancer.

Finally there are non-food applications. Several genetic modifications of crop plants have been designed solely for non-food products, such as industrial oils and biofuels. Some GM applications such as pest resistance and herbicide tolerance are applied to crops used for animal fodder, but which are not fed directly to humans.

The following list is not exhaustive but gives examples of current applications and near-market research. The applications that are asterisked (*) are in commercial use abroad. None are grown commercially in the UK.

Modifications directly affecting the food content of crop plants

- Slow-ripening fruit with longer shelf-life.*
- Changes in fat, starch or protein content to increase nutritional value and/or improve baking and food-processing qualities.
- Increased vitamin, mineral, nutrient or antioxidant content, for example vitamin A rice.
- Modifications to post-harvest behaviour, for example, prevention of sprouting in potato tubers, thereby avoiding the need to treat with chemical anti-sprouting agents.
- Removal of allergens and other potentially toxic substances from certain foods.

Modifications affecting crop 'performance'

- Increased resistance to viral, bacterial and fungal diseases.*
- Increased resistance to environmental stress, especially cold, drought and salinity.
- Increased resistance to predation by pests, especially insect pests.*
- More rapid production of flowers (and hence seeds) leading to a shorter growing season.
- Prevention of premature shedding of seeds.

- Diversion of plant resources from general biomass to specific production of the desired product (usually seeds/fruit). This includes the development of short-stalked cereal varieties.

Several of these applications have the potential to reduce markedly the use of agrichemicals.

Modifications affecting crop husbandry

- Herbicide-tolerant crops leading to improved weed control, using less weed killers.
- Short-stalked varieties (as already mentioned above).

Crops for non-food use

- Enhancement of plants' ability to remove toxic chemicals from contaminated soils.
- Production of pharmaceutical products in plants, including edible vaccines.★
- Production of industrial and research chemicals in plants.★
- Modification of oils, fibres and other polymers to meet new industrial requirements.
- Production of bio-degradable plastics, detergents, etc., in a sustainable production system.
- Expression of chitinase in roses to provide resistance to fungal pathogens, and many other applications.

Scale of Growth of GM Crops Worldwide

The estimated global area of transgenic crops for 2000 was 44.2 million hectares or 109.2 million acres. This is the first year that the global area of transgenic crops has exceeded 100 million acres, twice the area of the United Kingdom. There was a dramatic increase from 1.7 million hectares in 1996 to 40 million hectares in 1999, but a smaller increase to 44.2 million hectares in 2000, with some transgenic crop types showing a decrease in acreage. Over the same period, the number of countries growing transgenic crops went from 6 to 13. Up to 85 per cent of transgenic crops were grown in industrial countries until 2000, but the proportion grown in developing countries has increased from 14 per cent in 1997 to 24 per cent in 2000.

In 2000, four countries grew 99 per cent of the global transgenic crop area. Two of these were industrial countries, USA and Canada, and two were developing countries, Argentina and China. As has been true since 1996, the USA had the largest transgenic crop area in 2000; 30.3 million hectares, followed by Argentina with 10 million hectares, Canada with 3 million hectares and China 0.5 million hectares. In the USA the increase of 1.6 million hectares in 2000 was a result of an increase in the area of transgenic soya, cotton and canola, but for the first time there was a drop in the area of transgenic corn grown. In Argentina, a gain of 3.3 million hectares in 2000 was a result of significant growth in the sowing of transgenic soya and corn and modest increases in cotton.

The dominant type of transgenic crop is herbicide-tolerant soya, some 58% of the global area of transgenic crops in 2000. Globally, transgenic soya occupied 26 million hectares in 2000, with transgenic corn in second place at 10.3 million hectares and transgenic cotton third at 5.3 million hectares and canola (oil seed) fourth at 2.8 million hectares.

In 2000, the global area of herbicide-tolerant soya is estimated to have increased by 4.2 million hectares, almost a 20% increase. Transgenic cotton also increased by 1.6 million hectares in 2000, mainly in the USA, where 72% of the cotton is now transgenic. In contrast, the transgenic corn area in 2000 is estimated to have decreased globally by about 800,000 hectares with the major decrease in the USA and some in Canada. Some attribute this to the fact that USA farmers concluded that the low infestation of corn by the European Corn Borer in 1999 did not merit the use of Bt corn in 2000. Others have suggested that this was due to the uncertainty of the markets for transgenic corn, especially with European consumers reacting against GM foodstuffs. On the other hand there have been significant increases in transgenic corn planting in Argentina as well as in South Africa. There was also a slight decrease in the global planting of transgenic canola in 2000 with all of the decrease in Canada.

Herbicide tolerance is the dominant genetic trait to be engineered with 74% of the area sown in 2000. Insect resistance is second, with Bt insecticide accounting for 19 per cent of crops. The remaining 7% are plants that are both herbicide tolerant and insect resistant, and this trend for 'stacked' genes is gaining an increasing share of the global transgenic market. Herbicide-tolerant soya is the dominant transgenic crop, and is grown commercially in six countries in 2000 – USA, Argentina, Canada, Mexico, Romania and Uruguay. The second most dominant crop was Bt maize.

In the early 1990s, many were sceptical that transgenic crops could deliver improved products and make an impact in the near term at the farm level. There was even more scepticism about the appropriateness of transgenic crops for countries of the developing world. The steady increase over the five years from 1996 to 2000 is seen by the industry as justifying the early claims and expectations. The fact that so many farmers in both industrial and developing countries have taken on transgenic crops is testimony to the confidence they have placed in these crops. It is too soon to say whether the fall off in the expansion of GM crops from 1999 to 2000 marks a change in the trend or is a temporary blip. The number of countries deciding to grow GM crops has increased, but the picture is quite complex because some countries have made decisions not to grow them.

3

Is it Wrong to do Genetic Modification?

Theological Reflections on Genetic Modification of Crops and Food

In this section we address the underlying theological question of whether genetic engineering represents an intrinsically right or wrong intervention in creation. Terms like 'tampering with nature', 'unnatural' and 'playing God' have tended to be used rather loosely. We examine some key biblical themes and Christian perspectives which provide a theological context. We explore our understanding of the earth as God's creation and the place of human intervention in and care for the creation, the significance of the Fall, and the balance between dominion and stewardship. We believe the Old Testament has something important to say to us concerning how we are to live within our environment. God deliberately located his people within a marginal environment in which cohesive social and religious life and responsibly managed human intervention was directly linked to ecological well-being. Whilst we now think more in terms of global co-operation, we believe the consequences of ignoring God's world order still apply. We therefore consider the role of science and the biblical principles against which our increasing capacities to transform the world must be evaluated. This biblical view of science is established against the contrasting contemporary secular trends of 'scientism' which venerates science, and 'neo-paganism' which rejects it. We then examine some passages and concepts in the Old Testament Law which relate to the land, agriculture and Sabbath, but set these within the New Testament perspectives of the incarnation and the redeeming work of Jesus Christ, and the future hope for creation.

So, do we have the skills, insight and foresight to manipulate the genetic structure of creation? We investigate the key criteria as follows:

Creation

'In the beginning, God created the heavens and the earth' (Gen. 1:1). The Bible declares the universe to be God's creation. It is not divine or autonomous or co-eternal with God, nor is it simply the entity commonly called 'nature'. The Bible also celebrates the goodness of creation. It is declared by God to be 'very good' (Gen. 1:31) and, despite its abuse by mankind, it remains so (1 Tim. 4:4). The goodness of creation is manifest in a variety of ways. These include the provision of food for mankind (Gen. 1:29) and animals (Gen. 1:30), the beauty of creation (Matt. 6:28-30), as well as its sheer variety (Ps. 104:24).

The creation narrative (Gen. 1) proclaims God creating order out of chaos. In Genesis 1:2: 'the earth was without form and void (Heb. *tohu vavohu*) and darkness was upon the face of the deep…' *Tohu vavohu* is a picture of total chaos,[24] and is the opposite of the order that characterises the work of God. The creation narrative can be portrayed as a divine triumph over the forces of chaos as symbolised by 'water' and the 'deep', a motif that recurs elsewhere in the Bible to express God's sovereignty.[25] The divine word ('And God said…' Gen. 1:3) brings order out of chaos by a process of division and separation that establishes a series of categories. Light is separated from

---

24 cf. Isa. 34:11 and Jer. 4:23.
25 See e.g. Ps. 89:9 'Thou dost rule the raging of the sea; when its waves rise, thou stillest them' and Ps. 93:4 'Mightier than the thunders of many waters, mightier than the waves of the sea, the LORD on high is mighty!' and c.f. Jesus stilling the storm, Lk. 8:22-25.

darkness (Gen. 1:3-5); 'water' above from 'water' below (Gen. 1:6-8); water from land (Gen. 1:9-10); and male from female (Gen. 1:27; 2:21-23).

Wisdom plays a major role in creation (Prov. 8). It involves recognising and upholding these boundaries. Proverbs 8:29 describes how God '...assigned to the sea its limit, so that the waters might not transgress his command, when he marked out the foundations of the earth'. Some theologians imply from this that whilst chaos, symbolised by the 'waters' and the 'sea', is held in check by God's command, the potential remains for a relapse back into chaos, as when the antediluvian disobedience blurred the boundaries (for example in Gen. 6:1-2) which led to the flood.[26] It has been suggested that this might indicate a general principle that order and prosperity are linked to preserving God's categories.

Everything in the rest of creation is assigned its proper place within this tripartite structure of land, sea and sky according to their various 'kinds'. The land produces 'vegetation... according to their various kinds' (Gen. 1:11-12) and, later, 'the living creature after his kind...' (Gen. 1:24), and similarly creatures of the sea and the sky (Gen. 1:21). The Creator establishes a proper place for everything and sets boundaries that facilitate the orderly development of the world.

Humankind occupies a special place. Genesis 1–2 sets out a hierarchy in the order of creation. God is the Creator and all of creation is dependent upon Him. Men and women too are creatures. As with the animals, God breathes into us the breath of life (Gen. 2:7), but we are also made in the image of God, and given dominion over the rest of creation (Gen. 1:29-30). We are far exalted above all creatures (Ps. 8:6-8) and the rest of the creation.

What does this mean in regard to intervention in God's creation, and to science and technology?

---

26 Compare the prologue to John's Gospel, esp. Jn. 1:5. Light enters the world, but the darkness remains.

Humanity and creation

Genesis 1 and 2 present two contrasting and complementary pictures. Genesis 1 describes the whole creation in elemental terms, and uses the strong language of *dominion* to describe human relations with the creation. Men and women are commanded by God to 'fill the earth and subdue it' and to 'rule over' fish, birds, livestock, creeping creatures, and 'all the earth' (Gen. 1:26-28). This stresses God's calling to humankind, made in God's image, to express God-given gifts and creativity in transforming the natural world. Thus we may say that to practice technology, in its broadest definition of the 'practical arts', is a fundamental part of being human. This is expressed in passing in numerous ways throughout the Old Testament – in agriculture, irrigation, building, mining, metallurgy, ship-building, and all kinds of skilled crafts. It even provides the first occasion in Scripture when people are specifically said to be filled with God's Spirit (Ex. 31:1-11).

Genesis 2 expresses a counterbalance to Genesis 1, by describing human relationship to creation within the subtler context of a garden we are to work and care for (Gen. 2:15). Cultivation implies work and intervention, but it places more emphasis on the horizontal dimension. We are *fellow* creatures, to balance the vertical dimension of dominion. The Mosaic laws take this further with limits in patterns of land and animal use, as discussed below.

Thus in the very creation ordinances we find a creative tension between notions of development and conservation, which does not have a simple resolution. God calls us to work to find a dynamic balance between intervention and care. This represents one of the balances or tensions of the Christian life on earth, like law and grace, unity and diversity, or free will and predestination. It is less familiar, because environmental issues have not figured prominently in the churches down the ages, but it is no less real. Indeed, in the present context, it is a crucial insight. Our basic attitudes towards genetic intervention may be shaped by which side of the balance we feel more drawn towards. We note that to the extent that individual Christians are inclined towards one or the other, they may tend to see genetic engineering as favourable or dubious.

The notion of dominion or rule is not, however, to be read as mere licence to dominate and exploit, to do 'whatever our hands find to do'. In 1967, the historian Lynn White famously asserted that western Christianity is a prime causal factor in the environmental crisis.[27] Many evangelicals pointed out that in Scripture, if sadly not often in practice, dominion is moderated and interpreted by concepts where more relational elements are emphasised.[28] Wenham sees our ruling as being 'commissioned to rule nature as a benevolent king, acting as God's representative over [their subjects] and treating them the same way as God who created them'.[29] In this sense one might argue that human beings are called to 'play God' in a positive sense. The popular sense of 'playing God' reflects more a Babel imagery, of human schemes done out of the prime motive of declaring autonomy and supplanting God's sovereignty (Gen. 11:4), as we shall see below.

The notion of *stewardship* is frequently used to try and give a biblical expression to the balance between intervention and care. It is one of the themes in the *Evangelical Declaration on the Care of Creation*. The concept of stewardship itself needs handling with care, however. Like 'sustainable development' or 'precaution', it is a complex notion which tends to be interpreted according to the particular spectacles we or our culture are wearing at that point in history, as Richard Bauckham has recently noted.[30] On the one hand it is 'top down' imagery, for which counterbalancing concepts like 'companionship' have long been suggested.[31] On the other, it tends to lack an

27 Lynn White Jr., The Historical Roots of the Environmental Crisis, *Science*, Vol. 155, pp. 1203-1207, 10 March 1967.

28 Francis Schaeffer, *Pollution and the Death of Man*, IVP, 1970.

29 Gordon Wenham, *Genesis 1-15*, Word Biblical Commentary, Vol.1, Waco, Texas: Word Books, 1987, p.33.

30 'Stewardship and Relationship', in R.J. Berry, (ed.). *The Care of Creation*, Leicester: IVP, 2000, pp.99-106.

31 Society, Religion and Technology Project, *While the Earth Endures*, Edinburgh: Society, Religion and Technology Project, 1985.

adequate theology for the proper place of technology. There are numerous interpretations as to what it means in practice. The ground assumption, however, is that we are to intervene, perhaps by 'improving', 'preserving' and 'protecting' nature, but within limits.

The place of science

Out of this perspective, Christianity encourages a positive role for the wise ethical use of science in society. Science springs out of our invitation from God to have a relationship with him and with our world. Adam naming the animals is an expression of this. We should not, therefore, set science and God as being, by definition, antagonistic. We need to distinguish science practised in relationship to God from science as a secular ideology. Modern science arose in the context of the Reformation which recovered the understanding that the world is not a fearful place full of hobgoblins but God's creation. As creation, it reveals God's glory and is also a sphere in which God commands humans to exercise responsible dominion, acting as stewards to work in it and care for its upkeep.

A Christian outlook presupposes an ongoing discovery of creation. Moreover, whilst the Bible presents creation as a finished work, it is not static like a museum exhibit. The command to 'be fruitful and multiply' means constant change in the living world, regardless of any human intervention. The fact that we ourselves are made in God's image encourages human creativity in imitation of God's creativity. It encourages a responsible intervention in creation. It has rightly been pointed out that Christians have often been interventionist historically. There is indeed a connection between a western interest in investigation and the Christian tradition, even if White exaggerated the point.[32] In one sense, it could be argued that Christians have a duty to make life better, or else they are allowing sin to conquer. Historically it is of course a

---

32 Lynn White (1967), op.cit.

fact that Christians have often been keen to embrace new technology.

The Fall

God's plan for humanity subsists in right relationships with God (Gen. 2:7) and with God's creation (Gen. 2:15). According to the biblical account of the Fall as a result of human sin, among the consequences of the rebellion in Eden are a broken relationship with God (Gen. 3:8), a broken relationship with other people (Gen. 3:15f), and a broken relationship with the ground (Gen. 3:17-19). The Fall removed the ability of mankind as a whole to care for the earth in the way that God intended. Indeed, creation as a whole now goes in 'fear and dread' of what rebellious humanity will do to it (Gen. 9:2,3). The Fall has resulted in a world that humanity finds more difficult and dangerous than it was before the Fall. The backdrop to GM– the world of nature and the food that is grown – is influenced by the effects of the Fall.

It is as a result of the Fall that agriculture became so difficult ('cursed is the ground because of you; in toil you shall eat of it all the days of your life; thorns and thistles it shall bring forth to you; and you shall eat the plants of the field. In the sweat of your face you shall eat bread till you return to the ground', Gen. 3:17-19). Humanity's relationship to the soil is henceforth characterised by struggle. Instead of the solidarity of creation which had existed between 'man' and 'the ground' (Gen. 2:7), we are now locked in a perpetual struggle with the earth. This alienation is expressed by the 'thorns and thistles' which also symbolise God's judgement[33] and self-defeat (Prov. 24:31). The perennial threat posed by the 'thorns and thistles' reflects humanity's failure to rule over and subdue the threat posed by the serpent. Where once we could have ruled, now we can never 'subdue' the earth in the sense that God intended. As humanity resisted God, so now the earth resists humanity. Having

---

33 cf. the ruined city of Isa. 34:13.

stepped beyond our created function and challenged the created order, our relationship with the created order remains deeply disturbed.

The primeval history (Gen. 3-11) can be seen as confirming mankind's fatal tendency to overreach the boundaries defined for him by the Creator God. The fratricide described in Genesis 4:1-16 breaches a further boundary. The ground 'drinks' human blood (Gen. 4:11), and the result, for Cain the agriculturist, is further alienation from the soil (Gen. 4:14). A yet more extreme transgression is the attempt to cross the boundaries between earth and heaven with the interbreeding of 'sons [traditionally understood as "angels"] of God' and the 'daughters of men' (Gen. 6:1-4).

Humanity's recurring tendency to breach God's order is restated in the Tower of Babel fiasco (Gen. 11:1-9). This pattern shows an innate human tendency to go beyond the limits set by God at creation and to embark on 'anti-God' projects fuelled by human aggrandisement. The Christian perspective on science set out above comes into opposition to two fundamental errors in secular thinking about science, both stemming from naturalistic thinking, which remove God from the picture in two opposite ways. One is the elevation of science to the status of an independent force, and the other a neo-pagan aversion to science.

Science as ideology

The rise of modern science also owed much to the Reformation's challenge to philosophical theory or ecclesiastical dogma as primary interpreters of the phenomena of God's creation. The spirit of inquiry, which Francis Bacon and others advocated, encouraged people out to discover the actual facts, but nevertheless, remaining within a Christian understanding of the world, seeking to read God's revelation in the creation. The seeds were also there, however, of a secular mastery of nature, with no regard for God or of any external limits on human activity. Scientific determinism arose out of the eighteenth century Enlightenment project as the sovereignty of human reason, and the vehicle of progress, 'emancipating' humankind from superstitious belief in God's action. It elevated science to a world view, often known as 'scientism'. Its pillars consisted of objectivity, reductionism, determinism, and autonomous human reason, each of which represented a departure from the concept of science in the service of God. At root it asserted that 'things are what science says they are' and nothing else. Many writers, like Michael Polanyi and Donald Mackay have long since exploded the philosophical fallacies of the view,[34, 35] but it still has some vociferous advocates.

Objectivity assumes that the observer is able dispassionately to observe and to interpret the phenomena of nature, standing outside his or her personal beliefs, values and preferences. Whereas this is a useful pragmatic approach, as an ideal it is misleading. As we observed in Chapter 1, all scientists have beliefs about the world, which to some extent influence what they do and say about their science. Science itself is said to be 'value-laden'. Wider judgements may be involved in such things as choosing what to research, how data are interpreted and what is reported to whom. The assumption that when it comes to genetic engineering it is science that sees clearly and all others are biased by their world views has been criticised by theology as the idolatry of science, and by the social sciences as power play. The attitude which dismisses other viewpoints on scientific discovery and endeavour has been one of the root causes of the gulf between lay people and scientists over issues like GM food. Theologically, human beings are not autonomous, but interdependent in a web of relationships under God, and needing to bring the whole counsel of God to bear on everything we do, science included.

Reductionism is the assumption that things are what they are in their most reduced form, like genes or atoms. It reduces

---

34 M. Polanyi, *Personal Knowledge*, London: Routledge, 1958.
35 D. M. Mackay, *The Clockwork Image*, London: IVP, 1974.

the material world, and potentially even humans, to mechanistic interpretations and ultimately to their functions. It encourages a purely functional understanding of plants, animals and the rest of God's creation, which can be manipulated at will, if we can sec a use for them. It views genetics with a precision which is not always justified, and its orientation is to focus on the desired function and not to look for unintended effects. It underplays the interconnectedness which ecological and theological perspectives bring. It leads to an arrogance which has been over-optimistic in presenting what genetic modification will achieve. If things go wrong, science will be able to provide a fix in any case. It assumes that the only risks of genetic modification which should concern us are those which can be subjected to a scientific test, whose results will tell the policy makers what to do. This naivety fails to appreciate the role which ethical values play in making risk policy. As we discuss in the next chapter, it also fatally misunderstands the gulf between such a science-based approach and what the ordinary person *perceives* about the risks of GM food, regardless of their scientific validity, just as it missed the same gulf over nuclear power 20 years before.

In order to practise science, scientists have to make certain assumptions about the universe which are beliefs usually taken from outside science. The autonomy of scientism tends to reject any fundamental principles, especially religious and moral ones, and tends to deny that it has any of its own. It reduces issues to functional matters of positive and negative consequences, assessed by experts in a 'cost-benefit' analysis. It has led to a failure to realise the fundamental issues behind the emergent public concerns over GM food.

Neo-pagan views of nature

The opposite challenge is neo-paganism. A God-centred view of creation contrasts profoundly with the widespread rising trend of neo-paganism, which views nature as quasi-divine instead of created, and not to be tampered with lest 'she' strike us back. Neo-paganism glorifies not God but 'Nature'. What 'Nature' does is always assumed to be right. The view of nature, however, tends to be static and fatalistic. It is suspicious of science. Human scientific intervention is considered increasingly dubious the more deeply we disturb the 'nature' goddess' ways. It is irreverent to 'interfere' with the realm of the deity, and intervention is almost bound to go wrong. The goal is to act in harmony with nature's patterns and to 'balance the energies' from the imbalance we have created by our disrespectful interventions. It borrows terminology like 'patterns' and 'energies' from science, but with very different meanings and usage from the rigours required by scientific investigation. While relatively few may actually hold to such explicit views, recent surveys show a noticeable trend in this direction in the wider concepts of what people mean by 'God', and the roles of nature and science.[36]

It is important for Christians to be wary of this major cultural trend, and to be 'transformed in their thinking, not squeezed by the world's mould'. We must distinguish between a Christian view of care for creation and a neo-pagan deification of it. Christian ethics sees fallen nature as a morally unreliable and ambiguous guide to what is right. It is difficult to define since God clearly intends humans to intervene. In Britain, little of what we see around us can be said to be 'natural' in any meaningful sense to be useful as a basis. As already suggested, there is a tendency for what is 'natural' to be equated to a romantic idea of 'what things were like in our childhood', or 'our grandfather's time', which undoubtedly was far from 'natural'! Nature can also be destructive. The medical imperative to combat disease could be said to be working against nature, in the neo-pagan view of it. Christians insist nature may reveal God but is not God. For example, in Genesis 1:16 the writer makes a theological point over against pagan creation myths – the sun is merely a 'large light', not something sacred in itself.

---

36 Church of Scotland, *Understanding the Times*, Edinburgh: St Andrew Press, 1996.

We are not to worship the created or the creature instead of the Creator (Rom. 1:25).

Christian creation care – not scientism, secular environmentalism or New Age

It is more accurate to see the biblical critique not of human intervention in itself, but of how it is disfigured by our autonomy and pride, desire for power, exploitation of the weak, and loss of a sense of God's wisdom. Neo-pagan views make the first mistake; scientism makes the latter. Science can be seen as an expression of human creativity and a direct reflection of our being made in God's image. Care for God's garden must guide, assess and inform technological intervention, without necessarily precluding it. Sadly, where Christians have seen neo-pagan influences, they have overreacted and counselled church members to avoid getting involved in any environmental work as being 'New Age'. This is a most serious error. Whatever may motivate others, we are called by God to care for his creation. We dishonour God if we simply avoid this call out of fear of spiritual contamination. Indeed the world is surprised at how little the churches have shown an example.

Whilst Christians may find common cause on some practical issues, Christian environmental concern has different roots and goals from the secular model. We have a wider and stronger basis than the cause-effect relationships and the rights and responsibilities of people and nature of the secular model. Our stewardship stems from the reality of a Creator who will require us to answer for our use of that which is entrusted to us.

We see a broken relationship with creation as the outcome first of a broken relationship with the Creator. The agricultural and ecological crisis is not just the result of poor 'planet management' or even 'environmental sin', but is firstly a spiritual issue, a consequence of the Fall. Hosea 4:1-3, Isaiah 24:4-6 and Haggai 1:7-11 are examples where the failure of farming and pollution of the environment were consequences of moral failure, transgression and idolatry.

The Bible portrays the earth and the whole creation as having an origin and a destiny, as opposed to contemporary environmentalism based on either a closed system of mere evolution or the cyclic notions of the east. It puts humanity's place within this wider process rather than simply causing the earth to degrade. There is an ultimate hope for the earth, which does not, in the end, depend only on humankind getting its act together. Rather it rests on God's intervening through the incarnation, death, resurrection and return of Jesus Christ, which secures salvation not only for humanity but for the whole created order. This cosmic message contrasts with the despair of some contemporary secular environmentalism. Christians are to express the first fruits of that future hope in our care for God's earth, but the final realisation of that hope is not in our hands, but God's.

With such a mandate and such a message, Christians everywhere are called to take up afresh the creation ordinance to care for God's creation.

Old Testament Law and Wisdom

In this section, we examine some Old Testament texts relating to agriculture, seeds, jubilee and Sabbath. The laws commanded to Moses at Mount Sinai were given to govern the Israelites' life in the Promised Land. One of the defining features of that existence was a new form of agriculture (Deut. 11:10–11). Not surprisingly, many of the laws related specifically to land and agriculture. Even a land 'flowing with milk and honey' (Ex. 3:8) could be laid waste (e.g. by war, Ex. 23:29, 30). The Israelites were to care for the land (Lev. 25:1–5), domesticated animals (Deut. 25:4) and wildlife (Deut. 22:6). The people of God were to be the agents of God's care and concern for the land. Ultimately, the land belonged only to him. Agriculture was invested with theological significance. The choices the Israelites made could result in either fertile or infertile land (Deut. 28). Indeed, when the prophets denounced Israel for her sin, this was sometimes portrayed in terms of unfruitful land (Hag. 1). Hosea 4:1-3 presents a picture of the land mourning because of Israel's unfaithfulness to God. Apocalyptic visions of destruction are likewise presented in these terms (Isa. 24:1-12; Jer. 4:23-28). Yet there is also restoration in the form of a renewed creation (Isa. 65:17-25).

In the Bible, law and wisdom are closely linked. They find complete expression in Christ who fulfils the Law (Matt. 5:17) and who is the wisdom of God (1 Cor. 1:24). Law and wisdom are also intimately bound up with creation. The Law expresses the order of creation (e.g. Ps. 19) whilst Proverbs 8 describes the part of wisdom in creation. True wisdom is a result of being related to God through the person and work of Christ. It is expressed by a truly human existence characterised by a mastery of life. In the context of human intervention, wisdom could be said to be typified by choices that promote order and roll back chaos (cf. Prov. 8:29). In what senses would genetic modification express order or chaos? There are two levels to this question. One is the fundamental sense of order or chaos within creation, and the other its consequences in human and ecological affairs. We discuss these further at the end of this chapter.

Several aspects of the Law are especially relevant to the question of genetic modification. These include the laws relating to 'mixing' species, the concern for social justice and the Sabbath.

Mixing kinds

The laws against mismating cattle, sowing two kinds of seed in a field and making a garment out of two kinds of material are found in Leviticus 19:19 and Deuteronomy 22:9-11. However, they do not take us very far for contemporary application purposes because the prohibitions have a primarily cultural referent. Their objective is principally to maintain Israel's distinctiveness vis à vis the rest of the nations. The issue is the cultic 'mismating' of Israelites with non-Israelites. It is similar to Paul's concerns in 2 Corinthians 6:14-17. It does not exclude the possibility that these laws also reflect the need to preserve natural distinctions; but this motivation is not explicit within the text. It is noted also that the use of mules in Scripture includes animal breeding which crosses species, as seen in the very significant case of Solomon riding David's mule as a sign of kingship (1 Kgs. 1).

Certainly there was to be no mixing of Israelite beliefs with those of the indigenous population. Canaanite nature worship was directed to the propitiation, appeasement and manipulation, through sympathetic magic, of the gods of the Canaanite pantheon. By contrast, Israelite agriculture and the religious practices associated with it were very different, centring on thanksgiving, praise and remembrance of God's good provision through the Creation. The Israelites are warned not to forget God, to fall into worshipping other gods, or to trust in human strength.

Justice

The Law also contains warnings about exploiting the poor (Deut. 24:14–15), and this is tied up with condemnations of unjust agricultural practices (Lev. 19:9–10). There are warnings against over-exploitation of the land, and exhortations about providing for the poor. Very often the two go together. Leviticus 19:9 states: 'When you reap the harvest of your land, you shall not reap your field to its very border, neither shall you gather the gleanings after your harvest. And you shall not strip your vineyard bare, neither shall you gather the fallen grapes of your vineyard; you shall leave them for the poor and for the sojourner: I am the LORD your God'.[37] The verses about not harvesting up to the edge of the field demonstrate a broader principle than merely that of refraining from pushing the exploitation of land to its maximum economic potential. Even more, we are encouraged to strive for righteousness and justice in ecological relationships. This serves as a restraint on the profit motive and for the imperative of economic and agricultural efficiency geared solely to outputs.

---

37 cf. Lev. 23:22 and Deut. 24:20–21.

We must ensure that the agricultural system has built-in ways of providing for the needs of the poor and disadvantaged in ways that preserve their dignity.[38] Moreover, those who had benefited from the yield of the land were to share the blessing of the land with others. Deuteronomy 24:19-22 commands: 'When you reap your harvest in your field, and have forgotten a sheaf in the field, you shall not go back to get it; it shall be for the sojourner, the fatherless, and the widow; that the LORD your God may bless you in all the work of your hands ... you shall remember that you were a slave in the land of Egypt; therefore I command you to do this.' This undercuts the strong human temptation to greed and hoarding in the presence of plenty, and is a restraint to the practices of gene patenting. Consciously or unconsciously, by the exercise of patent rights, protection of one's own interests may place much needed developments out of the reach of the poor.

Again, this command is linked to social justice because it follows on the command not to pervert the justice due to the sojourner, the fatherless and the widow – the epitome of those who have no influence (Deut. 24:17-18). The prophets speak out against social injustice on the basis of the Law. Ezekiel condemns those who destroy lives to get dishonest gain (Ezek. 22:27). Then, as now, the rich are seen as killing the poor. At times they do this by grasping so much of the deprived's livelihood that they starve to death (cf. Mic. 2:1-2). Isaiah[39] speaks of God looking for justice (Heb. *mishpat*) but finding only *mispah* (bloodshed) and looking for righteousness (*tsedakah*) but hearing a cry (*tse'akah*). Instead of meting out justice, the rich are 'murdering' the poor.

From such Scriptures, many criticisms might be levelled at GM foods with regard to the question of social justice. This reflects, for example, the inequities in power structures by which large companies can dictate terms in promoting GM technology, without regard for the concerns of ordinary people, rich

---

38 In the Old Testament the poor are given the privilege of labouring for their own needs, rather than receiving a handout.
39 Isa. 5:7.

or poor, North or South. The focus on corporate profit in the markets of the North, as against meeting human needs, does not sit well with the biblical emphasis on compassion for the losers. These aspects are of course common to many other agricultural and trade issues. Because GM crops and food fall into this wider context of structural injustices, it has to be asked whether GM is more likely to foster the problems, or whether GM can play a role in finding ways to avoid or even solve them. We discuss this further in Chapter 7.

Sabbath

Brakes on the exploitation of agriculture and the poor are central to the various Sabbath institutions. The weekly Sabbath, the Sabbath year and the Year of Jubilee have one common theme: the suspension of individual rights over others in remembrance of dependence upon God. Rights over others, and over animals are suspended on the seventh day (Ex. 21:8-11; 23:9-13; Deut. 5:12-15); rights over debtors and the land every seventh year (Lev. 25:1-7), whilst the Year of Jubilee provided for a full-scale *restitutio in integrum* with the return of the dispossessed to their ancestral property (Lev. 25:10). The Sabbath taught Israel that nothing in creation was absolutely her own. This is the logical conclusion of Genesis 1. The Sabbath institutions taught the Israelites restraint in relationships where power is unbalanced, and restraint in regard to the natural world. This awareness of the absence of absolute rights over land, other people or animals is central to the Old Testament concept of how the human race should exercise responsible 'dominion'. The potential for abuse of 'dominion' is recognised as an ever-present danger.

The Sabbath institutions highlight issues of power and control. Jesus later opposed the egocentric imposition of one's own will upon others (Matt. 20:25,26) and introduced a different outlook. Those with authority are there to serve, empowering and giving support to those for whom they are responsible. The focus is directed towards those who are served and requires the humility described in Philippians 2:5-8. Its result is to set people free before God, rather than to try and dominate them.

Incarnation and Redemption

This brings us to the New Testament which at first sight appears to have little to say about creation. There is certainly a change of emphasis from the Old Testament insofar as the new people of God (1 Pet. 2:9) are no longer a particular race associated with a particular tract of land. However, the incarnation radically asserts the goodness of creation. God the Son becomes fully human (Jn. 1:1-4; Col. 1:15-20). The goodness of creation is also affirmed in Jesus' life and ministry. His miracles such as healing, feeding or stilling the storm can further be seen as restoring the goodness of creation corrupted and distorted by sin. Moreover, whilst sin has breached our relationship with God and with creation, Christ has reconciled us to God through his death on the cross (Rom. 5:1-11). There is a cosmic dimension to the reconciling work of Christ (Col. 1:20; Rom. 8:18-25 and Rev. 21:1-4). All things, including the creation, will find their fulfilment in him (Eph. 1:10).

Through the Fall, human activity in creation was marred. Romans 8 portrays the creation subjected to frustration by human failure to live out our calling. Yet God's calling to work in creation has not been rescinded. It is reaffirmed in the incarnation of Jesus Christ, redeemed by his work on the cross, transformed by his resurrection, and will be fulfilled at his coming in glory. Romans 8 presents the creation groaning in the eschatological hope that *redeemed humanity* will play again the role towards creation which God intended. As we have noted, this supernatural hope sets Christian environmental concern in a different context from naturalistic secular models. This should find present expression in, amongst other ways, intervention in creation with a transformed mind, as a first fruit of this redemption, and as part of our 'reasonable worship'.

The resurrection also affirms the goodness of creation because it was a material, bodily, resurrection (Jn. 20:20, 22, 27). The world of creation is not abandoned. Rather, the resurrection marks the beginning of its transformation, commencing in this life. Creation is being restored on the way to wholeness (Heb. *shalom*). In Christ, something changes about nature and the physical world. For some this may mean that everything is free or for others that everything is sacred, and therefore we should be careful of how we treat creation. Either way, the ascension, glorification and imminent return of Christ incorporates redeemed creation into the eternal purposes of God.

Theological Conclusions

We have considered God's creation in the light of the biblical records. Humanity is made in God's image and called to exercise dominion over the rest of creation, but to show responsible stewardship in the exercise of that dominion. The goodness of creation must be respected and this means balancing use of its resources for our needs with respecting limits set by biblical principles. These include the recognition that we are finite and fallen, that we have duties towards our fellow creatures, and are ultimately answerable to Jesus Christ, who is both the author and goal of creation. What then are wise conclusions to draw about genetic modification in crops and food?

We recognise that a variety of theological emphases may be adopted regarding the general issue of the care of creation. These can lead to widely different interpretations of the basic theological question whether GM food would be seen as inherently wrong, or is acceptable in principle. Some stress the calling to have dominion over the earth, and support genetic modification; others stress care for the garden and consider it represents too radical an intervention in what God alone has created. Both scientific investigation and technological application are expressions of God's image, in creativity and in fulfilling our role vis à vis the rest of creation. The perspective of the Fall, however, leads many to be sceptical about the misuse of science, especially in the hands of unredeemed humanity – that we can mislead ourselves into believing we know more than we do. Some are inclined to reject genetic modification on those grounds. The theology of wisdom is here interpreted in terms of restraint concerning the things we know little about.

Others consider that this is inconsistent, because Scripture shows humans intervening in all kinds of ways, including radical transformations of creation, such as mining and creating new metal alloys. They argue that in the past Christians embraced technologies with such far reaching and uncertain consequences as the printing press, industrial chemistry, electricity, the car and the Internet, and ask what is so different about genetic engineering that we should make an exception in that case? Here the theology of wisdom stresses the wise application of human knowledge, that even in our finiteness and fallenness, God still calls us to act creatively knowing that we cannot know all the consequences in advance.

Some cite specific scriptural passages as suggesting that in switching genes across species that do not normally mate, we have infringed a barrier set up in the separation into kinds in Genesis 1, or the particular prohibitions in the Mosaic Law. Is genetic modification blurring one of the fundamental distinctions? Others doubt that it is valid exegesis to identify the creation of plants and animals 'according to their kinds' with a notion of strict barriers, particularly given that the scientific concept of species is not something fixed. Boundaries may often be blurred. We note that there is a danger of Christians unwittingly borrowing from neo-pagan sources firstly, the notion that God's creation is static, pristine or sacred, or secondly, the superstitious attitude which suggests that we should not 'tamper with nature'. Neither of these approaches accords with Scripture. For example, our analysis of the texts in Leviticus 19 and Deuteronomy 22 about not mixing seeds and breeds suggests that these ought not to be regarded as general creation ordinances forbidding any cross species mixing. They seem more to relate to Old Covenant commands which emphasise Israel's ritual and genetic separateness from the Canaanite peoples, which have been superseded by the finished work of Christ, in the light of which Paul declares all foods to be clean.

We have thus found little substantial biblical evidence with which to support an *intrinsic* objection to genetic engineering as such. More searching biblical doubts about GM food may be found in the associated area of social justice, where power structures and commercial and materialist values lead us to express considerable concern. We therefore devote significant attention to these aspects in chapters on agriculture in the UK and the developing world. Some of these issues are of course not specific to genetic modification, and may be shared with 'conventional' agriculture. As we have examined GM food issues in their social and economic context, we find ourselves asking the potentially more disturbing and radical question as to how far our own Christian attitudes and lifestyles have been influenced by the cult of Western consumerism and the philosophy of the market.

At all times we should allow creation to point to the one who created it. Christians must seek ways of celebrating the goodness of creation as well as seeking to preserve it. At the same time, a realistic view of sin means that we can never be complacent about our relationship with creation. Mistakes are easily made, and constant vigilance is required. We are charged to care for God's world, and for many people it is the potential risks involved with genetic modification which present the main problem, rather than an intrinsic objection to GM as such. Risk is a complex area, with important theological issues of its own, and we have devoted the next chapter to it.

4

Is Genetic Modification too Risky?

Environmental and Health Risks of GM Crops

In any discussion of GM foods, it is not long before the question of risk is raised. Once raised, it is apt to pervade our thinking. For this reason, it is important to step back and ask some fundamental questions about the nature of risk as Christians perceive it, and as a biblical perspective would view it – which are not necessarily the same thing. In this chapter we examine the nature of risk, and summarise what are seen as some of the risks associated with GM crops and food, offering some comment on both health and environmental effects. We then consider a theological perspective on risk and precaution, and see how this might throw light on the risks of genetic modification.

GM in the Context of a Risk Society

There has been a significant trend over recent years in UK society to be both more aware of risk and more averse to it. Risk has almost become an alternative way of interpreting the world. Instead of the optimism of the decades immediately following the Second World War, based on an Enlightenment model of progress through knowledge and science, things are perceived increasingly in terms of their risk. This is seen in the business or financial world, personal life and relationships, and especially science and technology. The cause of such a general change of attitude has been the awareness that, for all that science and technology contributes many good things, the trust we put in science and the official assurances of government, industry or academics, some things have gone seriously wrong.[40] Events such as thalidomide, DDT, the ozone hole, Chernobyl, and global warming, have all eroded the confidence in science which was virtually universal, say, in the 1950s. A culture of blame and litigation has also emerged, which has made people and institutions more wary of being held responsible for their actions.

How then does a transformed Christian mind respond to these trends? Are we to regard novel technology as part of God's calling to humanity, or as something to be fearful of, opposed to God's wisdom? Is GM food part of God's calling to 'work the garden', or an improper risk which fallen humanity is taking at its peril? What is the mind of Christ on matters of risk?

By inclination, some people are more averse to risk, others are more risk takers. Some are more optimistic, others more pessimistic about the consequences of things – seeing a glass of water as half full or half empty. It is also easy to scare people about something which is unfamiliar, especially over food. With headlines like 'Frankenstein Foods' the media have been responsible for creating a powerful imagery of GM food as something fearful and irresponsible. The BSE crisis had already engendered a mood of distrust in scientific intervention in food production, and in our ability to regulate it properly. Into such a climate, it is hardly surprising that the public has fairly easily been persuaded to see GM food as very risky. Was this manipulation or the voicing of valid concerns? How seriously should Christians regard the presumption implied by the slogan, that in genetic engineering scientists are doing something inherently hazardous which may shoot off in some unexpected direction to produce a monster of a problem?

---

40 House of Lords (2000), *Science and Society*, op. cit.

We focus in this chapter on the two concerns most commonly voiced, namely that the use of genetic modification in connection with food might cause some serious problem either for human health or the environment. Before looking at these specific cases, we first identify some key questions for Christians which lie behind issues of genetic modification and risk.

- What are the criteria for assessing what is wise stewardship or caring companionship towards God's creation in general or of the particular species involved?
- Do we have the skills, insight and foresight needed to manipulate such deep matters as the genetic structure of creation? How different is this from any other human intervention? Does our fallen nature make any *special* difference in the case of GM food?
- Given our human finiteness, there is always uncertainty about any human intervention in God's creation. Since human activity therefore always entails certain elements of risk, how far should we be precautionary? What does appropriate precaution entail?
- We have only partial knowledge about the effects GM crops will have on the complex ecological systems. Does GM therefore rank as a misuse of human capacities and responsibilities, or as part of God's calling to subdue the earth and fill it?
- Do GM crops pose risks to health or the environment that are significantly greater than those posed by conventional breeding or organic methods of agriculture?

Understanding Risk

There is no such thing as absolute safety. Everything that humans do, or don't do, carries an element of risk. We can think of risk as having three parts. The first is the size of the hazard. How big would the consequences be, if it did happen? The second is the probability. How likely is it, or how frequently would it happen? Both of these can be calculated scientifically, to varying degrees of accuracy, to give a value for the risk which can be compared with the values of other risks. But this is not the full story. We may perceive a risk very differently from the answer that a risk assessor may calculate. We can play down a relatively high risk like driving a car when tired, yet be very fearful about a much lower risk of radiation from mobile phone masts. We can know about dire consequences of global warming yet still not cut down on our car use, because we don't want the inconvenience. There are many value, social and personal factors associated with these risk assessments we make every day. We place a weighting or emphasis on certain sorts of risks. For example, people tend to be more averse to large rare accidents than smaller ones that happen more often. Insidious risks are also more feared than ones we can see. We also distinguish between the voluntary risks we might choose to take ourselves – like rock climbing, driving a car or smoking – and risks that are imposed by others or by society as a whole – like passive smoking, other people's driving, or global warming. Some argue that humans prefer a certain level of risk, and if life gets too predictable, we go out and find new risks to take![41]

Technology can be seen as a kind of invisible social contract.[42] In return for certain benefits, we may be prepared to accept a certain level of risk from it, provided it meets certain criteria:

- *Familiarity:* how familiar the activity is to our own experience; how much we understand it.
- *Comparison:* whether something like it has gone wrong before, or has proved reliable.
- *Control:* how much we feel in control of the risk.

41 John Adams, *Risk*, London: University College London Press, 1995.
42 D.M. Bruce, "Playing Dice with Creation?", in *The Reordering of Nature: Theology and the New Genetics*, C. Deane-Drummond and B. Szerszynski (with Robin Grove-White), Edinburgh: T. & T. Clark, 2002.

- *Trust:* how much we trust those responsible; whether we share their aims and motives, or feel that their interests may cloud their judgement over risk evaluation.
- *Immediacy:* whether the impact is immediate but potentially avoidable, or hard to detect and therefore to be aware of.
- *Frequency:* if every fifth aeroplane crashed, we wouldn't fly; since it is very rare, most people take the risk.
- *Magnitude:* we tend to be more averse to acute events with large consequences than the same number of harms over a longer time or by a more chronic effect.
- *Benefit:* whether we stand to have a clear personal benefit from it. Mobile phones and GM foods represent contrasting examples. Few people see much to be gained from improving the production efficiency of large producers or the shares of multinational seed companies. But if GM offered a marked improved nutrition or quality of food, attitudes might be different.

What are seen as the Health Risks from GM Foods?

The first question most people ask is about any possible effects on human health. How do we know that the genetic modification, whether or not it directly affects the food content of the crop, does not have unexpected side effects? It is for this reason that all new food crop varieties whether conventionally bred or produced by GM techniques are subject to food safety checks. One problem is that the evaluation tends to take place in committees, and not all the data may necessarily be in the public domain. It is therefore easy to imagine all kinds of risks, but the present state of affairs was summed up at a landmark OECD international conference on GM food safety in Edinburgh in March 2000. In the presence of a large number of experts from all over the world, both proponents and opponents, no delegate could cite a substantiated case where the genetic modification of a foodstuff has been proved to be the cause of an adverse health effect on humans.[43]

Examples which have not been substantiated

One example that has sometimes been cited by campaigners and in media coverage is the case of the amino acid tryptophan produced by genetically engineered bacteria, and used as a dietary supplement.[44] As such, it escapes the rules in the USA that govern the safety of both foods and drugs, and here the problems start. Tryptophan may be produced by bacteria that have been genetically modified to make much more than they need for their own metabolism. Some of the excess is converted by the bacteria into a by-product that is harmless to the bacterial cells but poisonous, even lethal, to mammals. The tryptophan must therefore be properly purified. Several batches of tryptophan were delivered to the USA from a company in Japan which, in order to cut costs, had omitted some steps from the purification procedure. Because they also failed to carry out proper quality control on the product, the toxic by-products remained with the tryptophan. The results were horrifying. Over 30 people died and many more suffered from a crippling muscular disease. Despite much that has been portrayed in the media, this tragic case was not caused by genetic modification but conventional toxins.[45]

One of the catalysts of the GM food crisis was the controversy over the experiments of Dr Pusztai, on feeding rats with raw potatoes genetically modified to carry the toxic snowdrop lectin gene. Whilst deleterious effects were shown in the gut of

43 OECD Edinburgh Conference 'GM Food Safety: Facts, Uncertainties and Assessment', Edinburgh, 28 February to 1 March 2000.
44 D. Gershon, Genetic engineering: tryptophan under suspicion, *Nature*, Vol. 346, p. 786, 1990.
45 R. H. Hill, S. P. Caudill, R. M. Philen, S. L. Bailey, W. D. Flanders, W. Driskell, M. Kamb, L. L. Needham and E. J. Sampson, *Contaminants in L-tryptophan associated with eosinophilia-myalgia-syndrome,* Archives of Environmental Contamination and Toxicology, Vol. 25, pp.134–142, 1993.

the rats, the interpretation has been very widely criticised, along with Pusztai's experimental method. The claim that this was due to a section of promoter DNA taken from the cauliflower mosaic virus, used widely in many GM applications processes, was rejected by a formal Royal Society investigation.[46] The work has since been repeated by two groups who found no effect. Thus it seems that the experiments proved little, but at the time conspiracy theories abounded, doubts were sown in the public mind about the safety of GM foods, and the damage was done.

Some health risks under discussion

A more plausible concern is that if genes are added to food which were not normally found in the human food chain some of these might result in harmful effects on humans. One example is that a foreign protein derived from an added gene might cause an allergic reaction. It led, for example, to a company abandoning its experiments to add a brazil nut gene to soya, since brazil nuts are allergens for some people. It was the company itself that had predicted the possibility, and found it to be the case. The regulatory system requires that all novel foods are screened for known allergens, but it is impossible to exclude an unpredictable combination of allergens which might provoke an extreme reaction in a few susceptible people.[47] Such risks exist already with ordinary food, and are not likely to be significantly worse with GM food, given that they will have received especially close screening with just such questions in mind.

Another concern is in the use of regulatory genes in the process of genetic engineering. One example is the use of an antibiotic-resistant gene as a marker during the production, for

46 The Royal Society, *Review of data on possible toxicity of GM potatoes,* Ref. 11/99, The Royal Society, London, 1999.

47 C. Bindslev-Jensen, 'Food Allergy and GMOs', paper at OECD conference, *Assessing the Safety of GM Food*, Edinburgh, 28 February – 2 March 2000.

instance, of transgenic tomatoes or maize. In 1996, the Advisory Committee on Novel Foods and Processes (ACNFP) objected to the inclusion in Novartis' *Bt*-resistant maize of a marker gene which confers resistance to the antibiotic ampicillin. Although there was no physical evidence, there was a theoretical risk that this resistance could be transferred into human gut bacteria if the maize was used for animal feed in an unprocessed form, and the animals were fed to humans. This was overruled by an EC committee which considered the risk but saw it as negligible. Subsequently ACNFP have been vindicated for adopting a precautionary approach, because there is now a general acceptance in the industry to avoid such genes in future applications. Indeed, the real question is why Novartis did not act in a precautionary fashion in the first place, since the gene was only a tool used during the genetic modification, and could then have been removed. This has to do with risk perception, as we will see shortly.

There are also some potential pathways into the food chain of both humans and animals of non-food GM crops, for instance in honey produced from the pollen of transgenic oil seed rape grown for industrial oils, or GM feed used for animals which we then eat. It is very unlikely that a serious health risk would be posed from the secondary effects of eating beef from animals fed on GM animal feed, or honey from bees which had obtained pollen from GM plants, but it is important to have investigated such potential pathways. Another pathway was illustrated in the USA when a particular type of GM maize, which had not yet been passed by the US regulatory authorities for animal consumption, ended up being mixed with maize used in making taco shells for human consumption. The risks themselves were small, and the maize may have in due course been passed for human consumption in any case, but the point is clear. Regulatory decisions which distinguish between GM and non-GM crops, food and animal feed, on whatever grounds, are pointless unless the necessary segregation regime is established at the same time. This implies using separate equipment and rigorous administrative control.

Mistakes can always be made. On the other hand, segregation has been practised for many years in non-GM agriculture, for example to separate oil seed rape used to make industrial oils from that producing oils for human consumption.

New generations of GM products are being developed to incorporate nutritional or therapeutic proteins designed to improve diet, enhance levels of certain chemicals said to assist in resistance to cancers or other diseases, or even administer vaccines in edible form. Since the desired active chemicals are by definition natural products, it would seem unlikely that the mere act of producing them in food by genetic modification presents a problem which would not arise if they were extracted and administered separately.

What do we make of health risks from GM food?

Whilst we are especially sensitive to matters of food safety, in reality we know remarkably little about the safety of ordinary food. It is therefore important to ask what is different about GM? As discussed in the science chapter, some have raised doubts that the very act of inserting a group of foreign genes into an organism – whether primary genes or promoters used to regulate them – may have unintended effects within the genome of the recipient organism. This remains highly speculative, however. We already eat large quantities of DNA in any fresh food. In general, none of this DNA is incorporated into the human genome, but is broken down in our digestive system, or in some cases is excreted from the body unchanged. Whilst one could never rule it out, it is implausible to suggest that molecular genetic modification is intrinsically much more likely to lead to weird and harmful genetic changes to humans than eating normal food. In general, health risks are not so far seen to pose overwhelming problems for GM food. The key questions of risk have much more to do with the environmental effects.

What are Seen as the Environmental Risks from GM Crops?

There are many inherent uncertainties over environmental effects because we know quite little about the complex web of ecological relationships above and below ground into which agricultural intervention is made, whether GM or not. We do not have a sound basic understanding of ecological complexities into which we are releasing these crops. The more different genes and products we release the more complex it will become. It is then a matter of emphasis and orientation of one's viewpoint. This is so within the scientific community itself. Laboratory scientists typically point out that the gene constructs they use are most unlikely to have crucial selective advantages if they were to become taken up by other varieties and species in the wild. Ecological scientists, on the other hand, take their precedence from unintended effects when foreign species have been introduced, accidentally or deliberately, into native habitats. We focus below on two of the main areas of concern – biodiversity loss and unintended gene flow.

Will GM Crops upset the biodiversity of the local ecology?

When measures are taken to combat pests and disease in crops, there are likely to be impacts on other species along the food chain, whichever form of agriculture is used. For example, efficiently removing a type of insect may affect the population of birds that feed on it in that ecosystem. Agriculture always involves a balance between promoting the desired crop and discouraging its competitors and pests. There is no doubt that intensive farming has led to a loss of biodiversity, however. Agrichemicals such as pesticides are a major cause of this. The widespread introduction of autumn sown cereal crops is another important cause.

Genetically modified crops are not necessarily worse than conventionally bred crops in their effect on biodiversity. Some non-GM agricultural insecticides are designed to kill *all* insects. Indeed GM crops modified to be resistant to particular insect pests by producing their own insecticides are intended to reduce significantly the use of the insecticide. However, there are also potential problems. Will beneficial insects also be harmed? There is indeed some evidence of increased mortality amongst lacewing flies, ladybirds and in the USA, milkweed butterflies. Some of these experiments were, however, conducted under artificial worst-case scenarios which may not fairly represent what actually happens out in the field. One recent report from the USA suggests that the experiments on milkweed butterflies were misleading in this respect.[48] Another problem which GM crops share with conventional agriculture is the potential build up of pests resistant to an insecticide. For example, the extensive use of crops genetically engineered to produce the naturally occurring *Bt* insecticide has led to measures in a number of countries to restrict the proportion of a field sown with the GM crop.

In view of such effects, it seems wise to conduct farm-scale trials with GM crops to assess the likely effects on biodiversity before any commercial use is considered. In moving from greenhouse trials to the open field, such trials always run a small but finite risk of causing an ecological problem themselves. To minimise any risk it is therefore important that sufficient research is done under contained conditions to eliminate problems beforehand.

Will the genes escape?

One risk is the unintended spread of a gene into the wild population of the plant, or into weed species, thereby producing more insect, disease or herbicide resistant forms of the weeds. There are various mechanisms by which this could occur. One is via 'volunteer' crops. For all that one year's genetically modified crop may have remained well contained, it may

---

48 L. C. H. Jesse and J. J. Obrycki, Field deposition of Bt transgenic corn pollen: lethal effects on the monarch butterfly, *Oecologia* Vol.125, pp. 241-248, 2000.

be impossible to avoid some plants reappearing in that field in the following year. This in effect constitutes a form of escape, if the field is now sown with a different crop. Another is by pollination, wind or insect-borne, which may travel considerable distances, albeit not in large quantities. The Government's Advisory Committee on Releases to the Environment (ACRE) has published a summary of current knowledge on gene flow.[49]

The genes inserted by GM techniques are not autonomous entities. They cannot thus 'escape' as individual genes and then invade other living organisms. The new genes are integrated stably into the crop plant chromosomes and form part of its genetic make-up. This is checked over several years during the evaluation of GM varieties, just as the stability of inheritance is checked for crop varieties produced by conventional breeding (with the exception of first generation hybrids). So the real question should be *Will the crop plant hybridise readily with wild species, and if so what will the consequences be?* Furthermore, this question must be applied to both conventionally bred and GM crops because there is no difference between the hybridisation potential of a conventionally bred variety and a GM variety of the same crop species. A recent ten year study by Crawley suggests that crop plants do not readily establish themselves in the wild and that GM crops are no better at doing so than non-GM crops.[50]

For most crop plants such out-crossing is not a possibility. For example, there is not a wild relative of maize in the European continent. However, of the crops grown on a large scale in the UK there is for sugar beet a possibility of crossing with wild beet, and for oil seed rape a chance of hybridisation with wild mustard and wild radish.

---

49 Advisory Committee on Releases to the Environment, *Gene Flow from Genetically Modified Crops*, ACRE Annual Report 7, Annex F, Department of Environmental, Food and Rural Affairs, March 2001.

50 M. J. Crawley, S. L. Brown, R. S. Hails, D. D. Kohn and M. Rees, Biotechnology: transgenic crops in natural habitats *Nature*, Vol. 409, pp. 682-683, 2001.

Oil seed rape presents a good example. It is one of the leading contenders for growing GM products in the UK, but it also happens to be more likely to spread its genetic material than some other species. Rape and its wild relatives are insect-pollinated, and the pollinators certainly fly far enough to distribute pollen (whether from genetically modified or conventionally bred plants) over several hundred metres. Studies at the Scottish Crop Research Institute suggest that it is almost inevitable that there would be some transfer of genetic material to wild relatives.[51]

Even so, cross-hybridisation between such species must be considered a very rare event, and there is no certainty that a hybrid would be fertile. In this context, the hitherto notorious 'terminator' technology could actually have a practical role to play in the UK as a means of stopping unwanted gene transfer.

The possible reasons for failure of such hybrids to establish themselves is that firstly they may be less 'fit' than the parental species,[52] and secondly that they are in any case very rare. For example, in populations of wild radish growing in the vicinity of oil seed rape crops, the frequency of hybrids was less than one in one hundred thousand.[53] In populations of wild *Brassica* species in areas where oil seed rape is extensively grown, one hybrid plant was detected in 16,000 km^2.[54] However, even the rare occurrence of such hybrids between rape and its

51 Y.M. Charters, A. Robertson, G.R. Squire, 'Investigation of Feral Oil Seed Rape Populations', DETR Research Report No. 12, Department of Environment, Transport and the Regions, (London: 1999).

52 T. P. Hauser, R. B. Jorgensen and H. Ostergard, Fitness of backcross and F-2 hybrids between weedy Brassica rapa and oilseed rape (Brassica napus). *Heredity*, Vol. 81, pp. 436–443, 1998.

53 A. M. Chevre, F. Eber, H. Darmency, A. Fleury, H. Picault, J. C. Letanneur and M. Renard, Assessment of interspecific hybridization between transgenic oilseed rape and wild radish under normal agronomic conditions. *Theoretical and Applied Genetics*, Vol.100, pp.1233–1239, 2000.

54 I. J. Davenport, M. J. Wilkinson, D. C. Mason, Y. M. Charters, A. F. Jones, J. Allainguillaume, H. T. Butler and A. F. Raybould, Quantifying gene movement from oilseed rape to its wild relatives using remote sensing. *International Journal of Remote Sensing*, Vol. 21, pp.3567–3573, 2000.

wild relatives means that oil seed rape genes can find their way into the wild plant genomes if the hybrids themselves then backcross with the wild parent. This can of course only happen if the hybrids are fertile.

Nevertheless, the possibility of gene escape by out-crossing should be considered for every new crop variety with a careful analysis of the possible consequences, if any, of such an out-crossing. Thus it is wise to conduct field trials of any proposed GM crop that has the potential to out-cross. We note, however, that whilst GM herbicide-tolerant rape is subject to several years pre-commercial environmental evaluation, a conventionally bred herbicide-tolerant rape may be introduced without any environmental evaluation. This raises an important point – to be clear about what is the real concern in such an example. Is the problem genetic modification, or is it the concept of herbicide-tolerant crops, regardless of whether they are made by GM or 'non-GM' means?

The key question remains, however, not the mere fact of potential transfer, but does it matter in ecological terms? In the recent case of cross contamination between GM and non-GM oil seed varieties, English Nature have pointed out that, if normal crop rotation is practised in any given field, it probably will not matter if some genetically modified volunteers are present from the previous year's different crop species.

The GM pollen will certainly escape into the surrounding countryside. It may land on the stigmas of native plants like wild mustard or wild radish, and it may pollinate them. But the chance is very low, and the chance of a viable hybrid developing is lower still. If one or two per cent of the rape seed crop contains GM material, then we are considering something many orders of magnitude less than that when it comes to the chance of the modified gene getting into another species. And even if it does, it will not matter, because the herbicide-tolerant gene confers no advantage on that species, so it will die out very quickly. There could be a problem with 'volunteer' rape seed growing the following year, plants originating from seed which went astray at the time of sowing. Rape seeds are like small greasy ball bearings, and they escape quite easily. So they could perpetuate the gene flow. The answer is to spray the volunteers with another herbicide next spring, and then there should be no problem.[55]

What do we make of the environmental risks of GM crops?

As with health risks, this brief description cannot do justice to the complexity of the issues, which would require a book to itself. We have quoted some examples, partly to illustrate the issues, but also to point out that, while there are environmental risks, all is not necessarily as bad as sometimes appears in some of the more sensationalist reporting. There remains the possibility that eventually some GM crop or combination of crops may produce an unintended variety that does persist to a significant degree in the environment. It is important, however, to put this in the context of the ecological risks from unintended consequences down the ages in traditional agriculture, and more recently from chemically intensive forms. It may be argued that the uncertainties in selective breeding are at least as great as those from inserting individual genes. As with GM health risks, we must ask in what sense do GM crops present a greater risk than conventional crops? At present it is too early to conclude whether current GM applications do indeed consistently reduce herbicide or pesticide use. Not surprisingly this appears to depend on local conditions.

Farm-scale trials have generated much controversy. For some, the remote risk that they themselves would cause ecological damage renders them unacceptable, but we would approve the general principle of trials, to seek to resolve at least some of the issues. Doubts have been raised about their validity, but 'contained' experiments would be even less representative. It would be wrong to imply that GM crops only present environmental risks. Environmental agencies like English Nature and the Royal Society for the Protection of Birds, who have expressed much concern about the present GM crops under

---

55 BBC On-line Science and Technology pages, *Polluted Pollen's 'Limited Impact'*, 18 May 2000, http://news.bbc.co.uk/hi/english/uk_politics/newsid_753000/753586.stm

consideration, have nonetheless suggested that GM crops targeted more specifically to environmental benefits might in some cases be able to reduce impacts. The picture of GM risk is therefore a complex one, which does not give a simple 'yes' or 'no' answer, and needs to be seen against a wider biblical view of the world.

Theological Reflection: God, Faith and Precaution

Risk in creation

The natural world abounds in things which are hazardous to humankind, and indeed to other life forms also. For the human body to live and flourish it must be kept within certain limits – we are not designed to withstand certain forces, chemical environments and temperatures. The same applies to all other living things. God did not design humans to be able to fly or survive under water. This implies that such things as rivers, seas and cliffs present naturally occurring risks to human safety in the very way God created us and the natural world. There are also risks from external physical events, such as being overwhelmed by the effects of an earthquake, hurricane, flood or a volcanic eruption. Risk would appear to be something intrinsic in our present experience of living in God's creation.

The context we now see for risk is one disfigured by the Fall and our broken relationship with God, and by the spoiled relationships in every other sphere of life. In this state, we may fail to recognise risks that are there, or we may exaggerate them out of all proportion. Human beings may fall into new and greater areas of risk by the disrupted patterns of life we pursue in rebellion against our creator. But it seems that risk, as such, was there already. We should therefore resist the current notion widely projected by our culture that what is perceived as 'natural' must inevitably and infallibly be safer than the results of human intervention. There is a sense in which the restoration of fellowship with God involves faith rather than the proven certainty of sight. As Paul affirms, 'We live by faith, not by sight.'[56] At the very heart of Christianity, the final proof and vindication of our faith is a future event, not given to us until Christ returns. Paul expresses it starkly, 'If only for this life we have hoped in Christ, we are to be pitied more than all men.'[57]

There is also risk involved in creativity. This throws an important light therefore on the human activity we call 'technology'. Technology is one outworking of the creation of men and women in God's image, and God's command to work and care for God's creation, to fill the earth and subdue it.[58] The very act of doing this implies some measure of risk. It is in the nature of human creativity that until one has done a creative act, and maybe not for a very long time afterwards, one cannot see all the ramifications of doing it. It applies to all spheres of human creativity, and is an intrinsic part of the pattern which God has ordained for human life. It calls for humility but not necessarily aversion.

Risk awareness and illusions

One of the ironies of a technological society is that its very success at combating many of the hazards of life means that we have become less and less aware of the risky nature of creation. As a result, we do not appreciate how unusual and privileged it is to experience such a low level of natural threats to life. There has been a profound change in the immediacy of the experience of risk. To most earlier generations, it was all too obviously a normal part of life. Infant mortality figured in most families, mothers often died in childbirth, and life expectancy was short. Crop failure, famine and epidemic were common. For the present generation in Britain and for most industrialised countries,

---

56 2 Cor. 5:7.
57 1 Cor. 15:18.
58 Gen. 2:15, 1:28.

however, these hazards to existence have largely been brought under control, mostly through technological means. They have become remote to our experience. Now, when something goes wrong, it comes as a disruption from an assumed safe pattern of existence, not part of the normal expectations of life. We have come to regard this state of affairs as normal.

With this has come a shift in attitude, in which we have lost the notion of 'gift' – in this case the blessings of a society relatively free of natural risk – and turned it into something to which we have a 'right'. This is especially ironic when that attitude is applied to technology. There has also been a shift in *what* we are at risk from – from fears of natural disaster to those of human making. This has led to a present nostalgia in our culture to suggest that 'natural' is always better than human intervention, and the past safer than the future. The Bible makes plain that since humanity fell, there has never been any such 'golden age', and that what is 'natural' is no guarantee of safety. Have we then got so used to the idea of being relatively safe, that we have forgotten what it is like to be living on an inherently risky planet?

Risk is 'normal', in the sense that it is both part of everyday experience and a feature of the way in which God has created the universe for human beings made in his image. Therefore risk is not something thrust upon us rudely by scientists, politicians or others outside our immediate control, to disrupt our Arcadian existence. Our existence was already risky, because it is part of living. Part of the spirit of our age is the imputation that if a certain hazard is at all possible, then that is unacceptable, regardless of its improbability. This attitude is out of step with the way God has made the world. That level of certainty is not the basis on how life is to be lived. To demand it is to ask for a counsel of perfection. Nothing can be absolutely safe in this way. Indeed, seeking it may be regarded as a form of idolatry. Biblical hope rests finally on the promises and character of God, not on our attempts to build security based on externalities – as the rich fool discovered.[59]

---

59 Lk. 12:19-21.

The picture presented by the Bible suggests that security in life comes not from arranging external circumstances to minimise risk, but from walking with God through the chances and hazards of life. Paul speaks of hardships and pressures 'beyond our ability to endure', and learned that reliance was to be on God, not on himself or on externalities of circumstances. With no expectation that things were going to be any easier, he states, 'On him we have set our hope that he will continue to deliver us.'[60]

If this is so, then there are ultimately limits to precaution. Finite human beings will never have complete knowledge of any particular set of risks, and will therefore, to a greater or lesser degree always have to make a decision on limited data. One cannot continue for ever pleading uncertainty as a case for not doing something, because there always will be uncertainty. That is the very nature of reality as God has made it. No activity, however well tried and familiar, is free from risk. Indeed, for Christians this represents a real theological and apologetic question. Why do people worry so much about risk? Is there not a deeper question about mortality which it poses? Christians should be pointing people away from risk to relevant spiritual implications. We suggest that Christians should not react like the secular media which demand that 'if there is the least conceivable risk we should not do it'. This does not seem to be consistent with how God has made the world, encouraging exploration and discovery, albeit in a context of wisdom and prudent risk management.

Evaluating Genetic Risk

Against this context, the question for this or any technology is: in that case what is the balance of precaution and risk taking that is consistent with a Christian view of the world? What are we to make of the particular risks of genetically modified food

---

60 2 Cor 1:8-10.

to human health, society and the environment? It is not a simple case of calculating risks and saying, 'Yes that is acceptable', or 'No it's not'. Firstly, because we have not had a major disaster from GM on which to base calculations, GM risk assessment is forced to examine speculative 'what if?' cases, which are notoriously hard to calculate reliably. Secondly, it is impossible to demonstrate the absence of environmental risks in a laboratory. The point may come when a field trial is necessary. The very act of carrying out the experiment outside already poses the risk that is being assessed. Thirdly, as we have seen already, there are the many factors and value judgements which lead to the very differing perceptions of GM food risks.

One of these is the clash of ideologies between those who see the urgent need to abandon chemical-based intensive agriculture in favour of more 'sustainable' approaches, and who have chosen to locate GM as part of the intensive rather than the sustainable side of the equation. In some cases this is out of the conviction that to switch genes across species is inherently wrong, but in others it is felt that it is intrinsically too risky. Natural methods are seen as safer. As we have seen, this poses a problem for a Christian understanding of the world, since what is 'natural' is at best ambiguous in terms of its beneficence. The concept of genetic contamination if one pollen grain escaped presents an unrealistically pristine view of nature, which does not accord with the notion that God calls humans to intervene in and therefore, to a limited extent, to change nature.

Nonetheless, a proper Christian response must be duly sceptical of some of the claims of human mastery into which proponents of genetic modification have been prone. We have at best only a finite understanding of science, and we are inclined by our fallen nature to see what we want to see and not necessarily what is there. The scientific or commercial focus on the desired product can easily be oblivious to looking at wider effects and connections. The impressions of laser-precision targeted genetic insertion compared with selective breeding must be tempered by a statement of the uncertainties involved. At present, precision is achieved as much in the selection process to remove the pieces that have not worked, as in prescribing in advance exactly where a gene will go.

We have argued that the health risks are probably no worse than those we already run with existing foods, whether produced by organic or conventional methods. The evaluation of environmental risks depends more strongly on the context. The Church of Scotland argued against a general UK moratorium on GM crops, and asked for a more selective use of precaution.[61] This was partly because the risks among the crop species vary greatly. For example, oil seed rape has a number of species with which it might spread genes to a small extent, but the nearest relative to which GM maize could spread is across the Atlantic Ocean. The other reason was that, to be effective, a moratorium must specify precisely what information is needed, at the end of which a decision could then be made. In the present context it is very unlikely that after three or five years enough data will have accumulated from farm-scale trials and other research to remove all the uncertainties. At some point a decision has inevitably to be made on inadequate data. The question will still be about what level of precaution should be applied. This will depend on one's ethical evaluation of the factors involved in GM and not merely the question of risks.

Risk also needs to be considered in relation to the potential benefits. The perception that GM food offers only risks with no tangible benefits has been one of the major factors in consumers being very wary of GM foods. The Church of Scotland urged concentrating on 'applications that are restricted in scale, and which confer strong human or ecological benefits, and to set up long-term monitoring to see that there are no serious unintended effects over extended periods of time.'[62] In the

61 Church of Scotland, Reports to the General Assembly and Deliverances of the General Assembly 1999, *The Society, Religion and Technology Project Report on Genetically Modified Food*, pp. 20/93–20/103, and Board of National Mission Deliverances 42–45, p. 20/4.
62 Church of Scotland (1999), op. cit.

context of a UK ecology suffering from intensive agriculture, the question is whether any particular GM crop would be more likely to make things better or worse. Both sides are apt to leap on the latest piece of evidence and say 'There, I told you so!' in support of their particular standpoint. This may be equally true of environmentalists looking for any plausible evidence of the environmental problems they see as inevitable, and companies looking for any marketable signs of the success they predict. It might be concluded that to grow certain GM crops in the UK was inadvisable, but that others would be acceptable, provided there was a justifiable need.

On the other hand, an environmental risk which might seem questionable in a well-fed western European country, may not seem so important if one's overriding concern was simply feeding people. If a pest blighted a crop in a poor country and conventional breeding offered no solution, a pest-resistant GM crop might be welcomed, even though it might bring environmental risks. It depends on what is seen as more important. The same might apply to GM rice containing increased amounts of vitamin A, which might make a significant difference to nutrition and disease in parts of the world where the prospects of obtaining vitamin A by other agricultural, dietary or pharmaceutical routes looked unlikely.

Summary

This chapter has sought to set out some of the risks associated with GM crops and food, and to set them in context, both theologically and practically. The perception of risk can differ greatly from risk as seen by the scientific risk assessor. Risk must therefore be considered against a wider context of criteria than merely a scientific assessment. This includes the range of social, commercial, political, personal, ideological, media, and other factors – any one of which may determine an individual's view of risk. Throughout history we have made many uncontrolled interventions into our ecology, especially in the importing of foreign plants and animals. Each time a new eco-logical balance is reached, but never the same one as before. Occasionally things have gone wrong. Some Christians would argue the need to wait until we know more about genetic modification to be taking such risks. Others would say that although this sounds plausible at first sight, it is actually avoid-ing the issue by putting it off, and is a counsel of an unattainable perfection because it is impossible to know this type of risk in advance. They argue that UK society is probably overreacting to the risks which GM crops present to the environment or human health, if these are set in the context of the range of risks we already accept. Christians should be wary of irresponsibil-ity, on the one hand, but also wary of simply following the prevailing culture by refraining from acting if there is even the slightest risk. God calls human beings to act creatively in a context in which we are not given perfect knowledge or foresight.

5

Does the UK need GM Crops?

The Impact of GM on UK Agriculture

In the last two chapters we asked some fundamental theological questions about genetically modified crops and food, and also looked at their risks. The next three chapters examine some of the practical issues related to agriculture. Firstly, what does GM actually mean for the farmer and for UK agriculture in general? (Chapter 5). And what do we think about the claims of organic agriculture? (Chapter 6). And lastly whether GM lies too much in the power of large companies and the forces of globalisation? (Chapter 7).

Looking at the perspective of UK farming and farmers, once again we need to distinguish clearly between GM *crops* that are grown and GM *food* that is produced and eaten. Although there are many experimental GM crops grown at agricultural research centres and several well publicised farm-scale trials, at present no GM crops are grown commercially in the UK. Most of the GM food products that have been on the market in the UK were grown abroad. The tomatoes for the GM tomato paste were grown in the USA for the UK company Zeneca. GM soya and maize imported from North America have been widely used in processed foods, although the main supermarket chains have gone to considerable lengths to try to find 'non-GM' sources, with varying degrees of success. The claim to be 'GM free' may sometimes only be relative. One UK GM food remains on the market, however, in the sense that GM bacteria are used in the production of a vegetarian cheese – chymosin – as discussed in Chapter 2. Several other related GM products have been licensed by the UK regulatory authorities for production, including a GM yeast in beer making, but have never been marketed in practice.

The Farming Crisis

UK agriculture is undergoing a deep crisis. The blows caused by BSE, foot and mouth disease and other less publicised agronomic setbacks are set in a context of reduced subsidies from the EU and the challenge presented by globalisation. In a globalised market, where supermarkets are able to fill their shelves from almost anywhere in the world, it is proving increasingly difficult for UK farmers to compete against imported goods. The average annual income from farming for 1999 was estimated as £9,900, its lowest point for over 25 years. Its contribution to UK wealth has halved over 15 years. The trends are to reduce workforce and to cut overheads wherever possible, creating larger farms, more part-time working. This is adding to the individual farmer a sense of stress and isolation, which is not helped by a generally adverse media image. There is widespread evidence of high levels of stress among farmers and anxiety about the future of their livelihoods, and suicides have been increasing. Churches in rural areas from Caithness to Cornwall are becoming increasingly concerned about the lives and livelihoods of farmers in their communities. Large questions are being asked about what is the future for UK agriculture and rural life in general. How should we farm in the twenty-first century?

The state of UK agriculture

Declining:	Increasing:
Farm incomes	Part-time farming and employment in farming
Numbers of full-time farmers and farm workers	Average farm size
Number of farms	Public scrutiny of farming and consumer pressures
Contribution to GDP	Regulation and bureaucracy
	Organic farming and growing
	Interest in diversification

What role, if any, does genetic modification play in the future? UK farmers seem to be mixed in their attitude to GM crops. For some, they add another dimension to the crisis, presenting ethical and financial dilemmas, and increasing their vulnerability to public scrutiny. Others, however, regard GM crops as a potential solution for some of their present problems, some confident indeed that they herald another agricultural revolution. This would be dependent on consumer acceptance, and that is by no means certain.

Balancing the Costs and Benefits of GM Crops

What would GM mean to the farmer? At a practical level, the social, economic and environmental impacts of growing GM crops in the UK should translate into a new set of farm-level benefits and costs. The viability of commercial production would depend on obtaining a favourable balance between them. They are quite difficult to evaluate, and it is a matter of debate how far the necessary balance would be achieved in the short term.

The direct benefits to producers of GM crops would vary in detail, depending on the genetically modified trait. The first phase of applications have mainly been targeted at reducing costs by employing less chemical inputs, such as fertilisers, pesticides or herbicides, or using less cultivations. Other genetic modification aims to achieve direct higher yields to the farmer, or improved nutritional quality and product value to the consumer, thus potentially increasing returns. Other traits under experimental investigation include greater tolerance to environmental stresses. There are also a wide variety of potential non-food applications of genetic modification in crops, for example in producing a wider range of industrial oils, biofuels, biodegradable plastics, animal feed, and using modified plant viruses to produce vaccines and other pharmaceutical products in harvestable quantities within plant tissues.

Some doubts have been raised about the availability of data on farm-scale trials, and whether the genetically engineered trait will prove stable enough for the farmer.[63] One of the aims of the present farm-scale trials is to provide relevant data in the public domain. Studies in the USA and elsewhere in Europe indicate that once the initial crop evaluation has eliminated variability in the location of the gene, GM traits are as stable as the plant's native ones. The effectiveness of some genetically modified traits like herbicide tolerance may be sensitive if the plant is grown in conditions of environmental stress. In such situations, yields may be less than the conventional varieties. US farmers have, however, reported this is a small effect compared with other advantages from the crop, for example in reduced tillage of the soil and reduced chemical inputs.

In 1998, Lawton presented a UK farmer's more sceptical perspective on GM technology, commenting that 'we appear to be heading for marginal benefits while retaining all the risk, particularly the market risk caused by the apparent lack of public acceptability and real consumer resistance'.[64]

The direct costs of planting GM crops may include higher seed costs, licence fees, technology fees and possible additional

63 J.R. Franks, 'Genetically modified crops: some economic issues', *Farm Management*, 10, 1998, pp.107–117.
64 R. Lawton, 'Fields of dreams – a farmer's perspective', in *Old Crops in New Bottles?* Stoneleigh: Royal Agricultural Society of England, 1998, p.18.

costs arising from contractual requirements to use specified agrichemical formulations or not to save seed.[65] According to Franks,[66] Monsanto obliges any American farmer who plants its herbicide resistant soya beans to sign a contract which requires them to use only the same company's herbicide. They may also stipulate only one-season planting and company monitoring to ensure compliance. Others report that the company demands the right to inspect fields for up to three years after planting, has used private detectives to investigate farmers, and has prosecuted farmers who the company claim have breached its conditions.[67]

Data on the direct costs and benefits of growing GM crop varieties are available primarily from the US. For example, estimates have been made of increased margin to US producers from growing Monsanto's herbicide resistant soybeans instead of the conventional crop, based on data from various sources, including Monsanto. Seed costs could increase by 0–25%, offset by herbicide cost savings of $5–15 per acre and yield enhancements of 0–10%.[68] The fact that GM crops are so widely grown in the US suggests that US farmers find them commercially advantageous, but it remains to be seen what would be the UK experience, with a different balance of factors.

Among the differences are indirect costs. These seem likely to prompt greater concern among farmers than direct costs. These would arise primarily from environmental risks and food labelling requirements, and the associated regulations. For example, additional costs might arise if changes were needed in crop rotations or cropping patterns if it proved necessary to control genetically modified 'volunteers'. These are individual

---

65 Franks, (1998), op. cit.

66 Franks, (1998), op. cit.

67 A. Whitmore, 'Monsanto: corporate food gamble', in J. Madeley (ed.), *Hungry for Power: The impact of transnational corporations on food security,* London: UK Food Group, 1999.

68 R. M. Bennett, and A. Kitching, *Economic implications of imported genetically-modified soybean and maize livestock feed ingredients in the UK. Report of analyses,* Reading: Department of Agricultural and Food Economics, 2000.

GM plants from last season's crop that grew up with a different non-GM crop this season. For herbicide-resistant crops, if the resistance was transferred significantly to competitor plants or weeds, this would offset the expected decrease in herbicide use. In a worst case it could even increase herbicide use. There might be higher insurance costs to cover risks of gene movements to neighbours' crops. At present these various indirect costs are almost impossible to assess, and some may never arise. It has been suggested that in general UK farmers would incur larger indirect costs than their American counterparts, because of the tighter regulatory framework and the more complex nature of UK and European agricultural systems.[69]

More immediately, the production of GM crops like maize and soya *outside* the UK will impact on UK farmers if, as expected, these prove significantly cheaper. This affects the prices of products traded on the world market and the supply and prices of imported animal feeds, for example. The widespread uptake of GM crop varieties in North America and some other countries outside Europe, is prompting concern among some that the UK is getting 'left behind'. At a time when Britain's farmers face lower prices, increased regulation and the progressive removal of institutional support, some claim that competitors in the USA and Canada have a head start equivalent to five years' technological progress.[70] Using only non-GM soya and maize feed ingredients in the UK dairy, pig and poultry industries could increase costs substantially because of the widespread uptake of the GM varieties amongst overseas suppliers, and the costs of preserving the identity of non-GM feed soya and maize. If consumers were to bear these costs, it has been estimated that this would add 3–4% to the price of pork and 1% to the price of chicken.[71] On the other hand, the current

69 Franks, (1998), op. cit.

70 R. Turner, "Move forward now or be left behind", in *Old Crops in New Bottles?* Stoneleigh: Royal Agricultural Society of England, 1998.

71 R. M. Bennett, and A. Kitching, *Economic implications of imported genetically-modified soybean and maize livestock feed ingredients in the UK. Report of analyses.* Reading: Department of Agricultural and Food Economics, 2000.

demand by European supermarkets for non-GM produce has created niche markets for farmers in Brazil to grow non-GM equivalents.

Environmental Impact and the Farmer

The environmental impact of GM crops is of particular significance to the farmer and especially to Christian farmers. The rural environment in the UK continues to be threatened in many respects, and farming is in an ambiguous position. We are now more aware that farming plays a vital role in conserving the rural environment – what urban dwellers call the 'countryside' – but it also causes considerable negative environmental impact. These impacts, along with food safety, are a focus of public scrutiny and consumer pressure on farmers. This is a key factor in the divide between the urban mindset and the farming and rural community. There have been increased legislation and shifts in public expenditure aimed at 'greening' agriculture. The foot and mouth crisis has brought home a general, if belated, recognition that the maintenance of the rural environment depends to a considerable degree in maintaining a viable farming sector in the UK as a whole. In a period of decline in agriculture in food production, farmers may very likely need financial recognition for their largely unacknowledged role in environmental upkeep.

The fact that GM crops may impact negatively on the environment is a highly sensitive issue for farmers who grow them. This is partly because under the wrong circumstances, some of these impacts might 'kick back' in increased costs on the farm. The other aspect is that these issues have put some farmers in the front line in terms of popular pressure. Farms hosting GM trials have repeatedly been targeted by green activists, while consents for commercial production have been delayed pending the results of the same trials. There have been disturbing examples of pressure put on farmers and their families in their local communities by some green groups.

On the other hand, farmers and food producers who choose not to grow GM crops are concerned about the impacts on their production or the activities of their neighbours who do grow them. There have been well-publicised cases of organic farmers complaining at nearby farms hosting GM trials. One community in Highland Scotland has objected to the perceived taint of GM on the marketing image of the region's 'pure' Highland food. These are issues where local feelings can run high either way, and farmers involved may experience considerable pressure.

In this context, Christian farmers and others associated with agriculture are called to think and act over the matter of GM crops with integrity and in accordance with Biblical principles. This, of course, applies to farming in general, but the questions we have discussed above suggest the need for a wider critique of conventional agriculture and the consideration of the development of alternative, ethical and Christian approaches and practices. In Chapter 3 we discussed Biblical themes which provide a basis for Christian environmental responsibility and for environmentally responsible farming. Here is a considerable measure of common agenda between Christians and other environmentalists, but we also note some points of difference, as we shall also see in the following chapter.

GM Crops and the UK Farmer

The environmental impacts of GM crops may put farmers in the front line of public controversy. In terms of the development and control of GM technology, however, farmers are very much at the end of the line. Farmers may be central to the production of GM crops, but their voices are seldom the loudest, nor their actions the most powerful. The benefits, risks and costs seem likely to be unevenly distributed among suppliers, producers and consumers. Biotechnology companies may receive the 'lion's share' in many cases. Farmers face pressures from every side, from biotechnology companies, scientists, government, retailers, envi-

ronmentalists and consumers. All of these variously present
farmers with seemingly promising opportunities, or impose
constraints on their activities and decisions. One farmer, for example, expressed concern that 'GM' technology would result in
'ever more interference and control over our business, both
from the buyer and from the input end. Yet another body with
the power to enter and inspect.'[72] Others have a more welcoming view towards 'GM' technology. It would undoubtedly suit
some situations better than others.

At present there is a *de facto* moratorium on the commercial
use of GM crops in the UK. It remains to be seen whether the
current trials will lead to certain GM crops being accepted for
commercial production. The bioindustry hopes that the second
generation of applications, including so-called nutraceuticals
aimed at improving the nutritional or health benefits of food,
will seem to the public to offer the tangible benefits which the
first generation failed to provide. With current sensitivities towards food, however, it is uncertain how far such developments
would reverse the present adverse climate of public opinion regarding GM food. It may be that non-food applications of crop
genetic engineering, such as vaccines, biofuels and industrial
oils, provide the best opportunities to farmers.

In such circumstances, Christian farmers need a clear understanding of the issues. This not only in order to manage
their businesses with integrity, but also to care for their farming
neighbours and to be a prophetic voice in the agricultural industry. Equally, Christian consumers need to understand and
act in a way that recognises the role and experience of farming
people. This highlights the need for churches to tackle agricultural issues more explicitly and to help and support Christian
and other farmers in their decision making. They have an
important role to play in standing beside those working in the
farming world more generally, in the stresses and worries
under which many are often working. Rural churches especially can be an important witness.

---

72 Lawton (1998), p.18.

6

What Sort of Farming Should We Adopt?

Should we go Organic, GM, or What?

The primary purpose of farming is to produce food, though in most cases in the UK, the secondary purpose is to produce a satisfactory income for the farmer. Farming is founded in a pre-industrial, agrarian society, but at the end of the Second World War many people in the UK were nearly starving. There was not enough food to go round. This explains why there was food rationing. In the late forties, fifties and sixties there was a tremendous impetus to produce more home-grown food. A partnership between farmers, scientists and agricultural businesses resulted in a massive increase in yield per hectare of wheat, barley and other crops. By the late 1980s, imports of wheat for bread-making had fallen as newly-bred wheat varieties enabled bread to be produced from British-grown wheat. This was previously impossible due to the poor quality of the flour produced from native varieties. Yields increased greatly. High yields could be obtained from good agricultural land, whilst marginal land was no longer required for cultivation.

The Problem of Sustainability

However, these high yields resulted from high inputs of ferti-liser, pesticides and other chemicals. Arable farming in the UK had become high input, as well as high output. There is a strong argument that this type of farming could not be sustained over many years. More was being taken from the land than could be returned to it. Non-renewable fossil fuels are used in the manufacture of fertilisers and pesticides. High-yielding new varieties take many nutrients from the land which have to be replaced before the next crop is sown.

The main agricultural challenge for the 21st century is therefore how can UK farming remain sustainable? For some, the root of the present crisis is that modern agriculture is a by-product of the industrial revolution, dominated by a 'production and efficiency' mindset. Their solution is to develop a radical, alternative post-industrial agriculture. Others see the problem, but consider that present systems are capable of adaptation in much more sustainable ways, without necessarily calling for so complete a revolution as that entailed in the organic approach. A wide variety of arable farming systems have in fact been developed, many of which aim to be sustainable in one way or another. What we currently have in the UK is a continuum of systems in operation. At one end of the spectrum are the 'conventional' intensive, high input, high output systems which developed as we described above. At the other are the various organic farming methods. In between are several other management systems. We now examine three examples among this spectrum of options for agriculture.

High Input, High Output Intensive Farming

One extreme is to continue with the high input, high output system of arable farming of the 1980s. This puts in high fertiliser levels, protects seed from disease before planting with a pesticide, removes weed competition from the field with a herbicide, uses disease-resistant varieties, and sprays during the season with plant growth regulators, fungicides and insecticides (all types of pesticide). Often a farmer would drive through the crop five or six times in a growing season with a variety of chemical inputs. Intensive farming is obviously a very expensive process and suits larger units. The use of expensive specialised machinery adds to the cost. A combine harvester costs as much as a house in many parts of the UK. Obviously high yields of good quality crops are essential in this kind of system in order to recoup the high input costs and make a profit at the end of the year. Many farmers using this farming system would see GM crops as a better way of controlling pests, weeds and diseases with reduced levels of pesticide. They would not mind paying more for GM seed if they could save money by reducing pesticide inputs. They argue that they would be reducing environmental impact on the countryside at the same time. This type of farmer is happy to take part in GM trials.

This intensive system has had the greatest impact on the biodiversity of the countryside. Insects and weed seeds, which provide food for many birds, have been greatly reduced. Hedges have been removed to make room for larger machinery. It can also lead to high levels of waste and run-off into the water courses. Such effects as these, together with the general unsustainability of maintaining high inputs and dealing with the wastes, are seen by many Christians as incompatible with care for God's creation. Compared with the pre-war situation, intensive agriculture has led to very much higher yields, a wider diet and cheaper food, but its serious environmental costs have for many prompted a re-examination of our ways of practising agriculture.

Organic Agriculture

What is organic farming?

At the other end of the scale is organic farming. This has undoubtedly been the most publicised alternative approach in the UK, although it remains a relatively small part of the agricultural sector. In the UK it originated from the pioneering work in the 1930s and 1940s of Sir Albert Howard, Lady Eve Balfour and others. There are several approaches which fall under this heading, regulated by different bodies of which the most prominent is the Soil Association. Organic farming aims to produce good yield without artificial inputs. That means that all fertilisers should be natural, for instance using chicken dung pellets and manure, rather than inorganic nitrogen or phosphorus fertilisers. Pest, weed and disease control should also be as natural as possible. This makes use, for example, of mechanical weeding, and various ingenious systems for natural pest and disease resistance, pest and disease control, using careful planting patterns. Organic farmers are allowed to use some older pesticides under some circumstances, as a last resort, for example to control potato blight. Organic farmers are not permitted to grow GM crops or take part in the trials programmes.

The core of the approach was summed up in the phrase 'feed the soil and let the soil feed the plant'.[73] This focus on soil nutrition contrasted with farming based on soluble inorganic fertilisers, which directly feed the plant and, according to organic proponents, thereby impoverish the soil. Aside from its proscription of most inorganic fertilisers and agrichemicals, organic farming philosophy emphasises harmony with nature, soil health, energy conservation, food quality, farm animal welfare, and the avoidance of environmental pollution. Its proponents emphasise that it is not just about techniques but a whole interlinked system of farming in relation to the natural world. There are also other alternative forms of agricultures which have elements in common with organic farming but also distinct features. These include *biodynamic, Lemaire-Boucher, macrobiotic, mazdaznan* and *vegan* agricultures, and also permaculture.[74] Not all of these are currently practised in the UK.

---

73 E. Balfour, *The Living Soil*, London: Faber and Faber, 1943.
74 R. Boeringa, "Alternative methods of agriculture", in *Agriculture and Environment*, 5, 1980, pp. 23–108.

In general, it is harder to obtain a high predictable yield of good quality using this system. Yields may be quite low, and can vary considerably from year to year, which is why organic food frequently costs more in the supermarket. The farmer, however, must be able to receive a regular income from his or her crops. The demand for organic food and its production in the UK has risen rapidly in the last few years. At present the demand is such that many organic goods on supermarket shelves are flown in from abroad, which for the time being partly offsets its sustainable aims.

Many people are able to afford to buy this higher priced food and are happy to purchase it, because they feel that it is of a better quality than more conventionally produced food. At present there is little evidence one way or the other that organic food is necessarily safer or better for health, however, notwithstanding the basic claims. Such comparisons are actually very difficult to make. Some people buy it because they feel that it is more natural. It expresses the sense that recent intensive methods have perhaps gone too far in their intervention. As all farming systems are artificial when compared with the original landscape of Britain, the 'natural' argument is only relative and perhaps somewhat nostalgic. Organic farmers make use of good quality, up-to-date cultivars, and even allow some limited use of older pesticides in certain situations. Organic farming does not by definition imply 'pure' products, and is facing some safety questions of its own from the Food Standards Agency, for example over micro-organisms in manure and mycotoxins in grain. There is nonetheless an undoubted appeal for many people that organic food represents a more attractive and sustainable concept than food produced by intensive methods.

Organic agriculture has found a natural coherence with secular environmentalism as this has emerged over the past 30 years. Environmentalists generally regard it as being the best expression of the principles of environmental sustainability within the field of agriculture. The notion of sustainability has been somewhat devalued, being used to mean everything from continuous economic growth to radical green visions, but it represents one of several elements of environmentalism which have penetrated mainstream thinking and policy.[75] There is thus substantial organic research in mainstream centres such as the Scottish Agricultural College. It also provides a high quality niche market for farmers at a time of general agricultural depression. It represents one alternative – but not the only one, as we shall see later.

Organic farming and GM issues

The current standards laid down for organic agriculture proscribe GM crops, and the organic movement has become a vigorous opponent of GM foods and crops. Initially, GM crops were allowed and welcomed by some organic farming proponents and organisations, primarily in the USA, who saw them as facilitating an organic approach. In 1996, however, the council of the UK Soil Association debated genetic modification and made a decision to outlaw it as incompatible with organic and sustainable methods. There were three key arguments. One was ideological, drawing a sharp ethical line between selective breeding and genetic modification. The former was said to present a more natural way of agriculture, respecting the distinctions of species which have evolved, and seeing GM as disrupting inherent balances. Secondly, intervention by mixing across species was seen as inherently more risky. The risk argument was not primarily evidential. Indeed, it would be very difficult to prove scientifically, one way or the other. It was conceptual, associating one type of agricultural intervention with higher risks than another. The third reason was more practical, namely that organic farming does not need GM crops. It seeks to address the problems which GM crops address, but in other and, in their view, better ways.

---

75 DETR, *Quality of Life Counts. Indicators for a strategy for sustainable development for the United Kingdom*, London: Department of the Environment, Transport and the Regions, 1999.

Organic farmers have expressed concern that genetically modified crops may contaminate their crops. The Soil Association has, however, previously recognised that organic farmers cannot be responsible for the actions of nearby 'conventional' farmers. Hence a farmer's organic accreditation was not compromised by the drift of pesticide or fungicide from a neighbouring farm, or the possibility of fertilisation by pollen from the same crop species, conventionally grown on the neighbouring farm. Indeed, so-called contamination of the organic crop by a small percentage of non-organic material was accepted in the validation process. Similarly, a minimum threshhold for GM content in organic food is now recognised in EU standards.

A claim has been made that validation of organically grown sweetcorn would be threatened by pollination from a nearby field of genetically modified maize, derived from a different variety of the same species. Any planting of genetically modified crops must be approved by ACRE, the government's advisory committee on releases to the environment. The committee has to evaluate the rate and distance of pollen spread for all the crops it reviews. Maize is wind-pollinated. The wind can carry pollen grains considerable distances. The pollen, whether GM or not, dehydrates and becomes non-viable very quickly. Hence the ability to bring about a successful fertilisation falls off very quickly with distance. ACRE and plant breeders use these data to define planting distances, whether the varieties grown are conventionally bred or genetically modified. In this case, the net result is the possibility that a few seeds in a few cobs of 'organic' sweetcorn will have arisen from fertilisation by pollen from a GM maize. Under the normal terms of validation, the presence of seeds arising from pollination from a conventionally bred but non-organically grown neighbouring maize crop would not result in loss of the right to use the term 'organic' unless this rose above a certain level.

Some organic proponents argue that any level of mixing is too great and that GM must therefore be banned. This raises a general question about the rights of different groups to impose their own agendas on society at large. How far is society obliged to bow to one group who claims that the growing of GM crops represents a commercial threat, primarily because of the way they have chosen to define their own business? On the other hand, how far is another group justified in claiming it as their commercial right to have the opportunity to grow GM crops? Neither can presume an exclusive claim.

Finally, we would caution against the assumption that organic food is by definition safer than GM food and better for us. At present no one knows. The primary justification of the organic system lies in its overall approach rather than its results. As observed above, it carries some risks of its own. There is also a question of risk in relation to long-term food production. Various studies have questioned the ability of 'organic' to deliver enough food for 8-9 billion people, just as similar claims for GM food to 'feed the world' have been widely challenged. A precautionary approach might suggest not putting all our eggs into the organic basket.

Organic ideology and the spiritual dimension

The organic movement also has an ideological and even political dimension. Conford depicted it as a response to the inter-war dilemma of choice between fascism, communism and 'the decrepitude of bourgeois democracy'. Its proponents argued that a 'revivified rural life, based on the principles of husbandry, could provide an answer to the problems which beset Britain in the 1920s and 1930s. Work on the land would reduce unemployment; humus-grown food and open air tasks would improve standards of health; industry would be smaller-scale and geared to the needs of rural activities, and this would counteract centralisation and alienation ... Rural festivals and rituals connected with the cycle of seasons would reduce the need for mechanised entertainments of a jaded proletariat'.[76]

76 P. Conford, Introduction, in P. Conford, (ed.), *The Organic Tradition*, Bideford, Devon: Green Books, 1988, pp. 14-15.

Some of the pioneers were concerned also with social and spiritual values. Lady Balfour in particular wrote that 'when a new generation has arisen, taught to have a living faith in the Christian ideals, to value and conserve its soil, and to put service before comfort, then not only will our land have citizens worthy of it, but it will also be a land of happy contented people'.[77] Sherwell-Cooper established the 'God-fearing' Good Gardeners Association. Massingham denounced progress in the modern sense as the 'archetypal example of abstract thinking', and spoke out against the application to the land of the 'quantitative principles that govern modern man's urban life'. He saw industrial agriculture and the treatment of the farm as though it was just like any other business as the end products of a degenerative process that started with the Enclosures.[78] His ideal was to restore agriculture to its central place in English life, in a society of yeoman farmers and local craftsmen.[79]

A seminal paper by Lady Eve Balfour also reveals that although the organic movement has some Christian roots, it also drew on the anthroposophical ideas of Rudolf Steiner and other movements earlier last century.[80] It included concepts like energy flows and balances which are highly speculative and sit awkwardly both with scientific and biblical understandings. 'The energy manifesting in birth, growth, reproduction, death, decay and rebirth can only flow through channels composed of living cells, and when the flow is interrupted by inert matter [meaning in this case inorganic fertilisers] it can be short circuited with consequent damage to some parts of the food chain.' There are elements within the organic movement that we would identify with the types of beliefs commonly called 'New Age'. While it would be an exaggeration, however, to describe the organic movement 'New Age' as such, it

---

77 Balfour, *The Living Soil*.

78 H. J. Massingham, *Remembrance,* 1942. Quoted in: E. Abelson, (ed.), *A Mirror of England,* Bideford, Devon: Green Books, 1988, p.109.

79 Conford, *The Organic Tradition,* p.15.

80 E. Balfour, *Towards Sustainable Agriculture – The Living Soil*, IFOAM Conference, Dornach: Switzerland, 1977.

certainly has some advocates who would subscribe to such views. It shares an implicit idealism with secular ecological thinking about human solutions, which among other things fails to take sufficient account of human fallenness.

We have already noted in Chapters 1 and 3 the emergence of neo-pagan ideas in the culture at large, and warned against an implicit idealisation of nature and what is perceived as 'natural'. Equally, Christians ought not simply to run away from the organic movement for fear of spiritual contamination. Among its themes are respect for the natural world, care of the soil, health and wholeness, holism and connectedness, permanence, the recycling of wastes and responsibility for future generations. Christians find many of these terms resonate with their own expressions of care for God's creation. We may differ quite strongly about some of the movement's interpretation of them and the world view they fit into. But we prefer to see this as an opportunity to engage in dialogue concerning the spiritual dimension in agriculture and environment, rather than something to steer clear of. This is an area for evangelicals to reclaim lost ground, to listen with respect, and, as opportunity arises, to bear witness to the truth we find in Jesus Christ.

Indeed, it is also important to stress that not all who farm organically necessarily ascribe to it such ideals. It is a broad movement which includes many who would not subscribe to the complete ideology. For some, organic agriculture is simply a better way of farming. Lady Balfour's paper contains some valuable insights from 20 years of organic farming regarding soil properties, arable growth and animal husbandry, regardless of any particular philosophical interpretation. For others, it is merely a pragmatic choice to enter an expanding UK niche market. But there are many for whom the underlying values very definitely represent the motivation in seeking to demonstrate a valid and viable alternative. For them, the ideology is much more explicit and integral. It has a number of prominent advocates, including Prince Charles, who regards his approach to farming at Highgrove House as a demonstration of sustainable alternatives.

Should organic farming and other sustainable alternatives be espoused and advocated as a Christian approach? Should Christians farm and buy organic? Certainly this is the conclusion of some. The organic farming charity, Land Heritage, has a Christian foundation, and includes many Christians amongst its members and officers. It was at least one farmer's reading of the Bible that persuaded him to farm organically.[81] Others, however, question whether it is as 'good' as is often made out. Still others see it as a valid alternative system, but do not think it is actually necessary for farmers or consumers to go as far as its radical approach demands. They point to other alternative systems to intensive food production, which we must now also consider.

Integrated farming

Various methods and approaches to sustainable farming are being developed within the context of mainstream agriculture, and may prove to be of greater significance. Environmentalist, Jonathan Porritt, has predicted that 'the boundaries between what we now describe as 'organic', 'chemical' or GM, are likely to soften; whatever the descriptor, all production systems will be bound by the same discipline of sustainability'.[82]

Integrated farming represents a half-way house between very high input agriculture and organic farming. It aims 'to produce food profitably, whilst safeguarding the environment'.[83] It seeks to represent a balanced holistic approach to farming, which considers all aspects of food production. This includes the site and landscape, the nutrient status and structure of the soil, crop rotation, variety choice, pest, weed and

81 M. Huggins, "Western agriculture and Christian ethics", in S. P. Carruthers, and F. A. Miller, (eds.), *Crisis on the Family Farm: Ethics or Economics?* CAS Paper 28, Reading: Centre for Agricultural Strategy, 1996.

82 J. Porritt, *Playing Safe: Science and the Environment*, London: Thames & Hudson, 2000, p.90.

83 *Farmers Weekly*, 19 January 2001, p.5.

disease control, conservation, energy use, waste disposal, and also the management, auditing and monitoring of the entire farming process. Its scope is thus as wide as 'organic', but sets out from a different starting point. Attention is also paid to food safety, and care for the environment. It was originally developed by LEAF (Linking Farming And Environment) and in the Scottish TIBER project. The approach has now been developed on a wide front including by the Ministry of Agriculture in its programme for environmental management for agriculture, with extensive supporting materials. An EU-wide code for Integrated Farming has recently been published which aims to establish an EU-wide set of principles for sustainable agriculture.

Monitoring and auditing are a central part of this approach. For example, fungicides are only used to control plant diseases if there is a real risk of an epidemic developing. Plant varieties are carefully chosen with disease resistance. Both these measures result in a marked reduction in pesticide usage compared with a high input system, where insurance spraying is used. Soil is analysed for its nutrient status and a more exactly controlled amount of fertiliser is added to the soil for the particular crop being grown. A field can be analysed into zones, and different amounts may be added to different parts of the field. The net result is to reduce the run-off from surplus fertiliser. Particular attention is paid to the conservation of wildlife on the farm, including hedge and pond management, and woodland conservation. In this system, careful monitoring of all inputs aims to achieve high quality, high yield harvests with good profitability, because all inputs are used with precision and nothing wasted.

One attraction for many farmers is that this system claims to be more flexible than organic farming. It is not so much governed by a set of 'rules' as by 'approaches', and there need not be a prolonged conversion period. The aims of this approach combined with flexibility help to make this farming system a very attractive way forward, according to its proponents. For some Christians, this will not be going far enough in sustainability. For others, it represents a suitable alternative to high intensity agriculture on the one hand and some of the uncertainties of organic on the other. It also presents a solution to what some see as an unnecessary and perhaps somewhat political polarisation of the debate into either GM or 'organic'. Unlike 'organic', it leaves the door open to using GM crops if there was a point in doing so. Farmers using this system would be likely to consider GM crops on their merit, and would probably be willing to consider carrying out a GM crop trial if the advantages of the engineered trait in question were seen to be compatible with the aims of the system.

Conclusions about Alternative Farming Methods

Farming has always impacted on the environment. Intensive agriculture has increased the impact, and is high in its use of non-renewable energy, is highly interventionist and produces substantial wastes. Some consider it 'violent', reflecting the traditional view that farming is a *struggle* against the forces of nature. They advocate a paradigm shift to what they regard as more benign, 'non-violent', lower energy, information-rich and integrated farming, inspired by a belief in the possibility of farming in *harmony* with nature. These ideals find partial practical expression in organic farming. Others do not see the need for so radical an alternative, and look to see how conventional methods can be adapted to achieve results that might be equally good in terms of sustainability.

This analysis raises two important sets of questions. Firstly, do GM crops belong to the industrial-type of technology depicted in the first category above? In one sense GM crops are indeed 'interventionist' and some carry the risk of negative impacts on the environment. But they are also information rich and have the potential, at least, to reduce the use of non-renewable support energy in crop production and processing. It is also claimed they will potentially reduce the environmental impacts arising, for example, from conventional methods of weed, pest and disease control. Indeed, the organic

movement did not oppose GM crops until the last five years. It
is significant that Jonathan Porritt, whilst certainly expressing
deep concerns, recognised that 'there may well be a role for
certain GM crops' which 'may well prove to be of real benefit
to developing countries and to poor, subsistence farmers'.[84]
The resolution of this issue is likely to be more specific to particular crops and particular varieties, and subject to clearer
conclusions about environmental impacts.

The second question is whether GM crops enable the paradigm shift to be made to the second category of agricultural
practice. Is the ideal of 'harmony with nature' an attainable reality? Are its practical farming expressions ones which Christians should espouse and embrace? Whilst a Christian
perspective may advocate a measure of 'harmony' with nature,
this must be set against the outcomes of the Fall, and the neces-
sity of redemption. In response to human disobedience God
declares, 'Cursed is the ground for your sake; in toil you shall
eat from it all the days of your life. Both thorns and thistles it
shall bring forth for you...in the sweat of your face you shall eat
bread.'[85] The redemption for which the creation is 'groaning in
labour' will be fulfilled only when Christ returns.[86] There are
no secular utopias in the meantime. While we live in the 'in
between' times, we are called to show the first fruits of a renewed response to God's ordinance to all humanity to care for
his creation. This means a wise and responsible stewardship of
the earth's resources, as those who are accountable. But farming still remains 'hard work' and a struggle with 'thorns and
thistles'.

Christian environmental responsibility may find legitimate
expression in some alternative agricultures, even if the
underlying ideals and ideologies differ from those of Christian
exponents. Organic systems and integrated farming both
offer alternatives of sustainable agriculture, and a means of

84 Porritt, *Playing Safe,* p.89.
85 Gen. 3:18-19.
86 Rom. 8:22-25.

expressing Christian environmental responsibility and social justice to a greater extent than perhaps do conventional intensive approaches. In the case of organic agriculture, we advise that Christians should not 'buy into' the whole package including some of the ideological and quasi-religious elements. We need to remain critical of the approaches and movements where they deviate from Christian principles. While integrated farming may not have an explicit ideology, it too has its more subtle assumptions, which also need to be evaluated in a biblical light. The challenge is to develop approaches that enable Christians to farm in a way that reflects biblical principles yet remain in farming, and provide a basis for a stronger prophetic voice in the agricultural arena.

What are the options for the Christian consumer who is concerned to have a more ethical approach to food purchasing and consumption? For some, buying organic food will have strong attractions, but, as we have sought to explain, the picture is not so simple. Organic methods have their own problems too, and at present, while demand exceeds UK supply, a lot of organic produce has unfortunately to be flown in from abroad. In making their ethical judgements, Christians should be aware of these factors and of the fact that less radical alternatives also exist. These may satisfy many of the same concerns, and need not reject GM methods on principle. It would also be misleading to imply that simply by buying organic 'you've done your bit'. There are other important ethical considerations, such as sourcing fairly traded goods, buying locally, not shopping at out-of-town supermarkets, etc. Perhaps the central problem lies in polarising the question as GM or 'organic'? It tends to force a simple 'yes-or-no' answer, which fails to reflect the complex real situation about GM crops, either ethically or in practice.

7

Is GM just a Powerful Tool in the Hands of the Powerful?

The Power Structures Which Drive GM Crop Technology

Perhaps the most sustained criticism of GM technology applied to food and crops is that, whatever its merits in theory, in practice it is dominated by the goals, ambitions and philosophies of a small number of multinational seed and agrichemicals companies, operating in a globalised market. In this view, the context in which genetic modification operates inevitably means that any potential for good is misused in the interests of the powerful. Just as we were critical of some of the ideological assumptions of organic agriculture, so we must express our deep concern at the behaviour of the corporate sector and also at the way those in charge of policy in the public sector have mishandled their obligations to the public.

Genetic Modification and Globalisation

Genetic modification has come on the world stage at the same time as the complex processes which go under the general title of 'globalisation'. Two particularly significant influences are the treaties administered by the World Trade Organisation (WTO), and the reform of Common Agricultural Policy (CAP) within the European Union. The subsidies hitherto supporting, and to some extent driving, UK agriculture are under increasing pressure from trade rules. The need to control public expenditure and the enlargement of the Community are also likely to effect a reduction in subsidies. Although the EU will continue to seek to protect farming in some way, agriculture will increasingly be determined by activities on the world market.

Ironically GM crops mark a significant break in both trends. Within the EU the 1990 Directive governing the release of GM organisms has been in a state of virtual anarchy for several years, with each state pursuing its own interpretation. The consumer rejection of GM foods in the UK has led to changes across the whole EU. The food labelling legislation has been undercut by the trend to go 'GM frcc', and the much heralded directive governing GM patents has been challenged directly or indirectly by at least five member states. Fears that the USA and Canada would seek to force GM foods back into Europe by appealing to WTO trade rules was one of the key factors in the Seattle riots of 1999. Though they may hope for a change of climate, and seek to influence matters behind the scenes, for the time being few North American companies are realistically anticipating having a major GM market in Europe.

In the short term, the GM debate in the UK is likely to continue to be driven by the large retailers, and consumer and environmental lobbyists. Government influence is surprisingly small. The major UK supermarkets control more than 70% of all food retail, and profoundly influence all stages in the food chain. The recent history of GM crops and foods has drawn attention to the extent of this control, and to the power of the supermarkets in reflecting and shaping consumer opinion.

In the longer term, it remains to be seen whether the US public will turn against GM food, or whether the UK and Europe in general will accept it to a limited degree, or whether the status quo will continue. Much will depend on how far governments want to reform the WTO, and whether its rules will allow explicitly for national diversity over GM foods. The agenda may be increasingly set by the developing economies which can grow GM crops more cheaply, if they choose to do so.

Genetic Engineering and Corporations

Part of the phenomenon of globalisation is a shifting of power from governments and people to large transnational corporations, some of whose economic turnover exceeds that of many nation states. The rise of the global economy will make it even more difficult for governments to respond.[87] In this sense, GM crops are thoroughly globalised. For some years, a few large agribusinesses have been buying up seed producers on a worldwide basis, until they now control most GM crop development and the trade in transgenic seeds. Such a concentration of power is deeply worrying, especially when something as universal as food is concerned. Moreover, the potential benefits of GM technology have, to a significant extent, become subsidiary to the demands of market competition amongst these big actors.

Genetic modification can be an expensive technology. An initial discovery may perhaps be made on a relatively small academic grant, but that is only the beginning. To bring to market as a viable commercial product often involves heavy upfront financial investment and sustained high levels of skill in the development phase. Profits and dividends can be notoriously slow in coming. Many small biotechnology companies were floated on high expectations and amid much publicity, but later went into liquidation or were bought out by large multinationals. The costs of regulation and patent litigation can discourage smaller operators from gaining a market share. There are notable exceptions, but the trend is that GM technologies become concentrated in the largest companies, which are more able to bear the costs.

The agrichemical and seed industry had grown and prospered on the back of current agricultural practices, particularly through the demand created by the Green Revolution in farming, but was now in decline. To maintain commercial success and profitability it was only natural to seek out new

---

87 R. Scase, *Britain towards 2010,* London: DTI, 1999.

technologies to regenerate their important core business. They believed they had found this in crop genetic engineering. The main players and drivers of GM technology are thus publicly quoted chemical companies who have reinvented part of their core business through a spate of mergers amongst their own kind, and the purchasing of numerous seed companies. Intellectual property played a crucial role in patenting their own genetic technologies, and in strategic buy-outs of academics and smaller organisations who held patents on key technological steps which were crucial to gaining a monopoly in the desired product markets. Companies also lobbied hard to get the concept of gene and transgenic plant patenting universally accepted at international levels.

In doing so, they believed that biotechnology could offer new and exciting products that overcome the weaknesses of Green Revolution farming practices, as well as providing higher crop yields and other as yet unrealised benefits. The market for these GM products was seen as very lucrative. The current seed market in general is around US$23 billion. Some longer-term predictions have suggested that the GM market might increase from around 10% to a two-thirds share of an expected $30 billion seed market in the year 2010, compared to a market share of only 9% today. Monsanto was one of the leading players, and more than any other company saw a rich future from GM crops that would result from three waves of products:

> The first consists of GM crops which are resistant to insects and disease, or tolerant of herbicides. These will allow farmers to meet the growing demand for food… The second wave, due to begin in five years time, will see genetically induced 'quality traits' in food, such as high-fibre maize or high-starch potatoes some of which will help doctors fight disease. And in the third wave, plants will be used as environmentally friendly 'factories' to produce substances for human consumption.[88]

---

88 *The Observer*, London, 23 August 1998, citing Dan Verakis, Monsanto spokesman.

The primary goal of their business strategy was to convince farmers that they need the GM crops, in order to create a push for demand, while creating a corresponding 'pull' from the main markets in the developed world for GM food. This was to be done by convincing consumers of how biotechnology would solve world hunger or build a sustainable agricultural system. One zealous advocate of GM food proclaimed that, 'slowing its acceptance is a luxury our hungry world cannot afford.'[89] But the strategy was fatally flawed, as will be seen.

Consolidations through acquisitions, mergers and alliances have resulted in a concentration of corporate power that raises serious doubts about the ability of anyone to control them. These companies have gained a firm grip on the technology and its potential market, and how it will be developed. Furthermore, as a result of their activities, they have also given themselves other comparative advantages. One report summarises these as, 'a critical mass of R&D resources for funding long-term and speculative projects; economies of scale in relation to global markets; development costs that can be amortised over a long term; and expertise in marketing and distribution of seed'.[90]

This has happened at a time when public research institutions are increasingly being starved of funds. This has inevitably led to academic and independent research increasingly turning to such corporations for funding. With this trend comes a shift in overall priority of biotechnology research from publicly identified goals to private corporate agendas, which may be far from the needs, for example, of developing countries. With it also comes a corresponding lack of trust on the part of the UK public in the validity of the research.

89 *The Independent*, London, 25 July 1998, citing Business Access.
90 Clive James and Anatole Krattiger, *Biotechnology for Developing-Country Agriculture: Problems and Opportunities – The Role of the Private Sector*, Focus 2 Brief 4 of 10, October 1999, International Food Policy Research Institute (IFPRI), Washington.

GM food and crops have therefore become bound up in a wide and pervasive process of consolidation of power in the field of agricultural commodities, like seed and chemicals, and in the processes of food production and distribution. In chapter 2 we outlined the grandiose claims made for GM technology to deliver various benefits for the public good. The reality is that its priorities and direction have become increasingly dominated by the commercial goals of a relatively small number of private companies. The corporate agenda and demands of business can often strain against, and even override the preferred scientific approach. Competitive pressure tends to create the presumption to patent and commercialise as soon as possible. This has led to the charge that some applications are being rushed into production before proper research and trials are complete, often with little or no public debate, and without stepping back prudently to identify what is the best approach.

On the other hand, many scientists involved in these companies genuinely believe in the potential good the technology could bring, and amongst these are many sincere Christians. The companies and scientists defend themselves by pointing out that all the appropriate regulations were observed and their products passed the necessary tests, so they cannot be held responsible if certain groups of people do not like the results. Moreover, the politicians and their scientific advisors remain adamant that the scope of their regulations is appropriate. This raises a further question about the role of governments in such developments, especially in Europe.

Corporate and Government Power and the Consumer Rejection

Through the 1990s, the power structures that controlled biotechnology mitigated against an effective engagement with public opinion, and with wider values. The prevailing EU perspective on genetic engineering was well expressed in the preamble to the 1998 Directive on biotechnology patenting.[91] This brought into EU law the patenting of biological materials, genes and transgenic organisms which had been allowed by the European Patent Office for some years. The justification was almost entirely in terms of the benefits to the competitiveness of the European biotechnology research and industry. Similar goals were expressed by the UK and many other member state governments.

The EU accepted the view of US companies that GM crops which were not a problem for US consumers could be imported successfully into Europe. Monsanto made the mistake of assuming that farmers were their main customers, and that consumers would automatically buy the food produce, provided they had passed the due regulatory process. This worked in North America but was a gross miscalculation in Europe. In retrospect, their isolation and unawareness of what was going on in Europe was remarkable. They failed to appreciate the need to create a demand by convincing consumers of the benefits in more tangible terms than 'feeding the world'. The rhetoric failed to convince a sceptical Europe who saw potential health and environmental risks instead of improvements, the main benefits primarily going to North American producers, and little or nothing being done for the world's hungry. Monsanto's 1998 advertising campaign was a spectacular failure as far as the general public was concerned, but a leaked internal report still nurtured hopes that they had the 'political elite' (MPs and civil servants) on their side, and that all would be well.[92] The cynical disregard that emerges for the concerns and values of the ordinary people of Britain makes chilling reading. The spectacular downfall of Monsanto's approach in Europe, and its subsequent demise as a company, have become a case study in corporate failure.

---

91 European Commission, European Parliament and Council Directive on the Legal Protection of Biotechnological Inventions, EC 98/44, Brussels, 1998.
92 "Re. the British Test – the Fall 1998 Research", Monsanto internal report, 5 October 1998.

Meanwhile the EC was in difficulties over its stance on BSE and animal hormones, and was not prepared to risk the USA imposing trade sanctions. The UK Government was focused on deregulation to create faster markets for biotechnology products to compete with the USA.[93] The technical argument of 'substantial equivalence' was cited as the scientific basis for accepting most of the GM products into Europe. It was logical, but socially unacceptable, because only scientific challenges were permissible, thereby marginalising public opinion and values that were not expressed in scientific forms.

Indeed, opposition based on ethics or precaution was frequently dismissed by government and the scientific community as irrational because it was unscientific. It ignored various warnings from those who could see history repeating itself, comparing it with the history of the nuclear industry. A key lesson of the latter is that moral objections or adverse perceptions of risk cannot be reduced to questions of 'scientific evidence', and then be expected to receive public approval. To be advised that a food is acceptable because it contains no detectable traces of the foreign genes or proteins proved of little actual relevance. If people object on ethical grounds, or because of their environmental concerns about the way the original crops were grown, or because they distrust the regulators' judgement, test results do nothing to address their concerns. All three were central features in UK consumer attitudes.[94]

Not all companies behaved as badly. Zeneca (now Syngenta) carried out careful market research before sending their tomato paste into the shops, and heeded the result, which was that any GM food should be clearly labelled, even though the law did not require it. They, and other companies, warned Monsanto not to bring GM products on to the UK market without labelling. Monsanto and Novartis responded that to segregate GM soya and maize from non-GM, and

93 House of Lords Select Committee on Science and Technology (1993) 'Regulation of the United Kingdom Biotechnology Industry and Global Competitiveness', HL Paper 80 and HL Paper 80–1, HMSO: London.
94 ESRC (1999), op. cit.

label its products accordingly, was both impracticable and uneconomic. It was more truthful to say that American producers, the UK Government, or the EC, were simply not prepared to take seriously public views, in case they impacted on a commercial course of action which had already been decided upon. This failure to segregate and label was central to the rejection of GM food by the UK consumers in 1999, for whom the availability of choice was all-important. Sociological studies indicate that preferences which consumers may state in surveys are not necessarily acted out in practice, but not *having* any choice is anathema. It is now clear that segregation and traceability of commodities, though difficult, is always possible if there is a will to do so – as organic producers have demonstrated. New EC proposals in August 2001 mark a radical change in recommending the labelling of food as GM by its production process, not just its content backed by extensive traceability requirements. It will cost money, however, which raises the question whether it is a new GM product or an existing non-GM product which should rightly bear the segregation costs, or both.

Each of the parties involved has lived to regret what can only be termed arrogance and insensitivity. Monsanto was merged into a new company, named Pharmacia, and its besieged biotech business spun off as a subordinate company which has kept the name Monsanto and headquarters in St. Louis (USA), but as a shadow of its former self. Zeneca lost a successful product as a result of others' actions. The UK government has seen its policies frustrated by a widespread consumer rejection of GM foods, and the EU regulatory system in this area remains in serious difficulties.

The GM crisis has delivered a shock to government and industry alike in the unexpected power of consumer demand, at least where food is concerned, if the media and green NGOs see an issue big enough to push, and where there is a public sensitivity. All these came together over GM food. It has showed an ironic vulnerability of what had seemed all-powerful companies, if they abuse their position, as

Monsanto did, and also of government and EU policies, where they fail to appreciate deeper public values. It has seen the emergence of new power players in the supermarkets, the media, the organic movement and the resurgent green lobby groups. Each of these has its own agenda. Already some disturbing features are emerging that show, as with all revolutions, that the new power brokers in turn are not above suspicion. Wherever power is located, it has the potential for abuse. With a sanguine view of human nature, Christians should therefore be vigilant and raise questions about the behaviour, influence and accountability of the power holders, new and old, whoever they are.

Power in a Post-modern Context

We have described a picture of globalisation, corporate power and political focus on corporate, rather than public, interests as the context within which genetically modified food and crops have emerged onto the world scene. It would, however, be quite unjust to lay the concerns expressed above at the door of GM itself. These represent much larger structural and political-economic trends. But in such a context, GM crops can be argued to be 'a powerful tool in the hands of the powerful'.

Hans Kung observed that the model of 'progress' has been its own undoing in giving rise to unaccountable centres of power:

> The Enlightenment itself shattered some of its own basic assumptions. There was progress in scientific research in all areas, but where was the contemporary moral progress that would have prevented the misuse of science? A highly efficient macro-technology developed which has spanned the world, but the spiritual energy which could have brought under control the risks of technology did not develop to the same degree. There was an economy which expanded and operated worldwide, but where are the resources of ecology for combating the destruction of nature which is equally worldwide? In the course of a complex development, democracy has slowly been established in

many countries outside Europe. What has not been established is a morality that can work against the massive power interests of different men of power and power groups.[95]

In the absence of a moral consensus, the post-modern world lacks checks and balances to restrain undue use of power. Veith considers that 'Without moral absolutes, power becomes arbitrary. Since there is no basis for moral persuasion or rational argument, the side with the most power will win. Government becomes nothing more than the sheer exercise of unlimited power, restrained neither by law nor reason.'[96]

Some argue that in a post-modern world, truth is much less likely to be discovered than to be constructed. If this is the case, then we could expect the facts of an issue to be both of less consequence and more difficult to establish, and campaigning stances to be the most important aspects of debate. Something of this has been seen over the past two years in the politicisation of the GM food debate in the UK. It therefore becomes increasingly important to identify the motivations and assumptions of the various players and stakeholders. In the GM debate, there has been an ironic twist to the topography of power. Far from having had it all their own way, the traditional sources of power in industry and government have been undermined by unexpected expression of consumer power, and the very astute campaigning of networks of environmentalists, consumer and development groups.

To reject GM crops just because they are part of this corporate and governmental power structure would seem too simplistic an answer however. On this basis large swathes of contemporary life would fall. The question is whether there is any redemptive aspect which Christians can identify and help effect change. Biblical priorities emphasise justice, mercy, concern for the disadvantaged and the poor, care for God's

95 Hans Kung, quoted in: G. Cray, 'Postmodernism knows no commitments', *Relationships Foundation Seminar,* 18 November, 1995.

96 G. E. Veith, *Guide to Contemporary Culture*, Leicester: Crossway Books, 1994, p.159.

creation, and the sharing of God's common grace in both creation and human creativity amongst all people. Thus far the primary values which have driven GM products have been competition-driven agendas of the powerful. It remains to be seen whether other values can play a significant role, and especially the voices of the poor, as we shall discuss in the following chapter.

8

Is GM Needed in the Developing World?

Motives, Accountability and Priorities

For many years, the proponents of genetically modified crops took as their ultimate justification the claim that they were needed to feed the projected expansion of the world's population through the twenty-first century. The Green Revolution had brought food to millions, it was suggested, but selective breeding had limited potential to feed an additional 3 billion or so people. What was needed was a new technological method. This was the cue for the entry of GM food, stage centre. This ethical high ground claim has been much debased, but does a more modest version still have any credibility? On the other hand, as numerous opponents have maintained, is GM food the last thing the poor and hungry of the world need? We now examine the case for and against, in the context of food supply and agriculture in the developing world.

The Stark Reality

The following summary describes the situation in the developing countries which forms part of the setting for the GM debate. It portrays a scene of detrimental social and environmental changes that are occurring in the Developing and Third Worlds. A situation, as the next section affirms, that will only become worse as the world population continues to grow.

> Today, almost a billion people live in absolute poverty and suffer chronic hunger. Seventy per cent of these individuals are farmers – men, women and children – who eke out a living from small plots of poor soils, mainly in tropical environments that are increasingly prone to drought, flood, bushfires, and hurricanes. Crop yields in these areas are stagnant and epidemics of pests and weeds often ruin crops. Livestock suffer from parasitic diseases, some of which also affect humans. Inputs such as chemical fertilisers and pesticides are expensive, and the latter can affect the health of farm families, destroy wildlife, and contaminate water courses when used in excess. The only way families can grow more food and have a surplus for sale seems to be to clear more forest. Older children move to the city, where they, too, find it difficult to earn enough money to buy the food and medicine they need for themselves and their young children.[97]

The population of the world will increase substantially over the next few decades, and will undoubtedly exacerbate the situation just described. Analysts estimate that the world population will rise from approximately 6 billion today to perhaps 8.5 billion by the middle of the century.[98] By far the majority of this increase in population will occur in the world's poorest nations. If we are to avoid misery on a massive scale then food security for all must be paramount. Food security is the state in which all people at all times have access to enough safe and nutritious food to maintain a healthy and active life.[99] This key goal has remained both elusive and distant. What then can or should be done to reverse this situation and establish proper food security for the world populations?

---

97 Gabrielle J. Presley and John J. Doyle, *Biotechnology for Developing-Country Agriculture: Problems and Opportunities: Overview,* Focus 2: Brief 1 of 10, October 1999, International Food Policy Research Institute (IFPRI), Washington.

98 H. W. Kendall , R. Beacht, T. Eisner Gould, J. S. Schell, and M. S. Swaminathan, *Bioengineering of crops: report of the World Bank Panel on Transgenic Crops,* International Bank for Reconstruction and Development/World Bank: Washington DC., 1997; I. K. Vasil, Biotechnology and food security for the 21st Century: a real-world perspective, *Nature Biotechnology,* 16: 1998, pp. 399–400.

99 Panos Briefing, 'Greed or need? Genetically modified crops'. Panos Media Briefing No 30, October 1998.

Biblical Reflections

Several Biblical themes offer a challenging commentary on these issues. The first is God's passionate concern for justice and his judgement on those who oppress the poor. Whilst it is almost characteristic of the rich to oppress the poor (Jas. 2:6) and the 'poor will always be with you' (Mk. 14:7), it is offensive to God and a mark of a decadent society if the rich get richer and the poor become poorer. This was, for example, the condition of the society which the prophet Amos confronted and which provoked God's judgement (e.g., Amos 5:11-12). God 'looks for justice' (Is. 5:7), and social justice must be one of our first and central concerns in understanding and addressing the GM phenomenon.

There is a second but related issue here that highlights the need to be alert to the power structures of our times and what the powerful are doing. Revelation 13 points not only to an apocalyptic situation, but also to a recurrent theme in human society. It warns that a key aspect of the control that those in power may wish to exert is over basic human needs (vv. 16-17). If power over the food supply is being concentrated in the hands of smaller and smaller numbers of people and power groups, and if GM crops assist this process, then that is an issue of Christian concern.

The third power issue relates to knowledge and wisdom. The forbidden fruit was the fruit of the tree of knowledge, and one of the temptations was being 'desirable to make one wise'. The Tower of Babel account presents the basic human desire to become powerful, to become like God. Isaiah depicts a judgement that comes on all the earth, because 'they have transgressed the laws, changed the ordinances and broken the everlasting covenant' (Is. 24:5). Daniel 7 introduces a 'beast' who 'thinks to change the times and the law' (v. 25). The desire for power that is in the heart of humanity is, ultimately, the desire for power over life itself. As such, genetic technology applied to maize, soya and oilseed rape may seem fairly innocuous. But what it represents in control over food and livelihoods, however, is a serious concern when power is concentrated in the hands of a few who do not have the trust of society nor the requisite accountability to it. This is even more important in relation to developing countries, as we shall see.

Current Farming Practices – Do they Represent the Future?

The Green Revolution has been a source of endless debate, claim and counter-claim. According to whom one listens to, it has either provided millions with food security and transformed a country like India into a net food exporter, or alternatively it has devastated the lives of traditional farmers by importing western ideas and methods inappropriate to their culture, resources and land. Anecdotal evidence is cited in support, either way. In reality, there is probably no single overall robust conclusion. It would seem to have worked well in some situations and badly in others.

Where it has been successful, it has led to greater productivity and efficiency, through the introduction of large-scale, monoculture, mechanised, chemical-intensive farming, resulting in increased crop yields. It has been estimated that over the past two decades there has been an increase of 15 per cent in the amount of food available for the current world population, and many hail this as a success. But does this way of farming represent the future? Can it produce ever higher yields to feed a growing world population? Can it overcome the problems poor farmers face? The answers to these questions are hotly disputed.

Against such successes, the Green Revolution has also received much heavy criticism. Florence Wambugu of the African regional office of the International Service for the Acquisition of Agri-biotech Applications (ISAAA) argued that its performance has been inconsistent. It is claimed to have failed in Africa because it was an alien import from the West. Its chemical-intensive methods are accused of destroying the environment and the health of farmers in the developing countries. Among the problems cited are the use of greater quantities of water, herbicides and pesticides. The World Health Organisation estimates that at least three million people are poisoned by pesticides every year, and more than 200,000 die. It is estimated that up to 25 million agricultural workers are poisoned every year.[100] More chemical-intensive farming is seen as heading in the wrong direction.

There are also still millions of malnourished people living in the developing countries who have obviously not benefited. According to the UN's Food and Agriculture Organisation (FAO), one in seven people are chronically malnourished, including one in three children. Many people in developing countries, if they manage not to die of starvation, lack the necessary levels of proteins, vitamins, minerals and other micronutrients in their diet. Gordon Conway, president of the Rockefeller Foundation in New York, provided some examples at the recent OECD Edinburgh conference on GM food safety:

> About 100 million children suffer from vitamin A deficiency. They are more likely to develop infections and the severity of the infection is likely to be greater. Each year half a million go blind and some 2 million die as a result. Iron deficiency is also common. About 400 million women of childbearing age (15-49 years old) are afflicted by anaemia caused by iron deficiency. As a result they tend to produce stillborn or underweight children and are more likely to die in childbirth.[101]

The debate rages as to which claims are justified, but there are certain shortcomings that even the Green Revolution's supporters recognise need to be addressed. Yet the solutions themselves are controversial depending on which side is taken.

One approach depends on using technological advances made by biotechnology through genetic engineering to

---

100 WHO quoted in "The Facts on Pesticides", *New Internationalist*, May 2000, No. 323, p.19.
101 Gordon Conway, *Crop Biotechnology: Benefits, Risks and Ownership*, Paper presented to the OECD Conference, 'GM Food Safety: Facts, Uncertainties and Assessment', Edinburgh, – 28 February to 1 March 2000.

overcome past shortcomings and meet the needs of the future. It is argued that GM crops could eventually eliminate chemical inputs that harm the environment, provide better weed and insect control, higher productivity, higher nutritional benefits to millions who suffer from malnutrition and deficiency disorders, and flexible crop management. For example, a World Bank panel has estimated that transgenic technology can increase rice production in Asia by 10 to 25% in the next decade.[102] This is still based within the same paradigm standing behind the Green Revolution. The groups, farmers and companies who stand to win the most here are the same ones currently benefiting from the Green Revolution, although they claim it will help everyone. Again, this is either a good or bad thing depending on one's beliefs and interests.

The alternative approach calls for a paradigm shift. It argues that the Green Revolution and biotechnology are not the solution. Instead what is needed is the creation of a just and equitable system of food distribution, coupled with viable and productive small-farm agriculture based on land reform using principles of 'agroecology' that includes organic farming.

There are numerous others who stand in between. Some lean towards the alternative approach, but also appreciate the benefits biotechnology can bring to farming in the developing world. Some suspect the agendas of the biotech industry, and retain a cautious view. Still others believe they can work with them in a 'win-win' partnership to their country's advantage.

Do We Need GM or a More Just System of Food Distribution?

This is a key question in the debate. It is another hotly disputed topic as people consider the best way forward for creating

---

102 IFPRI report (Oct. 1999).

proper food security. Some would answer with a simple 'yes' to the question, whilst others may suggest 'yes, it would help immensely, but it is not the whole answer'.

Many opponents of the Green Revolution and biotechnology argue that genetic engineering in agriculture is not necessary, pointing out that there is enough food being produced now in the world to feed everyone with a nutritious and adequate diet. They claim that most people who go hungry or are malnourished do so primarily because they are denied access to food. A CornerHouse Briefing asserts that, 'a whole range of unjust and inequitable political and economic structures, especially those relating to land and trade, in combination with ecological degradation, marginalise poorer people and deprive them of the means to eat'. The briefing also adds that 'they starve because they do not have the money to buy food, or do not live in a country with a state welfare system'.[103] Many others echo this particular view. The main focus of GM technology, which is currently increasing the efficiency by which crops are grown in the American mid-west, is seen as largely irrelevant to feeding hungry people in Africa, if inequitable distribution is the fundamental problem. GM will also not be of any use in situations where the problem is less about distribution than the need for reforms in land ownership and tenure. Moreover, they argue, these GM technological 'fixes' are similar to putting a plaster over the problem only to find the patient is allergic to the glue used, and more importantly do nothing to prevent the hurt in the first place. In other words, they allege, biotechnology does nothing to address the underlying causes of hunger, and on the contrary is likely to exacerbate them. What is needed, it is argued, is for these underlying causes to be directly addressed and rectified without turning to biotechnology for help.

Many supporters of modern farming methods would agree that proper and just food distribution would play a vital role in

---

103 *The CornerHouse,* Briefing 10: 'Genetic Engineering and World Hunger', October 1998.

providing food security, but it will not solve the problem by itself. There is still an important need to ensure that modern farming methods can continue to increase yield outputs, particularly as available land becomes scarcer at a time when the world population is increasing. Likewise, some scientists such as Florence Wambugu, believe that transgenic crops are going to be crucial in achieving the necessary higher yields that places like Africa need, but admit that they alone will not solve all the problems. Other voices question whether it is possible to have a just system at present, and given this fact it is even more imperative that technological solutions are found to increase yields. Suman Sahai, president of the Indian NGO Gene Campaign, was highly critical at the OECD conference about the priorities of GM research and activities of the controlling companies. She also acknowledged that there were agricultural problems in pulse production, drought tolerance and disease which GM might help to address, if someone would carry out the necessary research. This was conditional, however. It must feed the poor, give local farmers a sustainable living, and avoid control or ownership by multinational corporations.[104]

Gordon Conway expresses a similar view when he asserts that he is not convinced that industrial nations, and those in the rest of the world who are benefiting from the current system, are ready to deal with the underlying structural issues in any serious way. Since the world is not about to engage in a massive redistribution of wealth, he maintains that the only way to help the majority of the world's poor, who tend to live in rural areas, is by increasing their income and food productivity by sustainable agricultural and natural resource development that involves the use of genetic engineering. In this light he believes:

> We need a new revolution – a doubly green revolution – that repeats the success of the old but in a manner that is environmentally friendly and much more equitable. This is going to take the application of modern ecology in such areas

104 Suman Sahai, *Concerns about GM Crops*, Paper presented to the OECD Conference 'GM Food Safety: Facts, Uncertainties and Assessment', Edinburgh, 28 February to 1 March 2000.

as integrated pest management and the development of sustainable agricultural systems. It is also going to need much greater participation in the development process by farmers themselves. But he also believes it is going to need
the application of modern biotechnology – to help raise yield ceilings, to produce crops resistant to drought, salinity, pests and diseases, and to produce new
crop products of greater nutritional values.[105]

Gordon Conway also takes issue with the fact that at present
powerful private corporations dictate and control the majority
of the work in the biotechnology field. What is needed, he
insists, is for it to be taken out of their hands and for it to be
placed within the public domain and targeted towards helping
the poorest nations. One could level the same criticism of this
suggestion as Conway does at the idea of transforming underlying structural problems that lead to hunger and poverty. A
fallen world needs a reality check on some of its aspirations, but
this is no reason for not trying. This issue will be taken up in
more detail later.

Professor Swaminathan, one of the world's most respected
agronomists and father of India's Green Revolution,[106] offers a
way forward. He argues that not only do we need to address the
very important issues concerning the underlying structural
causes of hunger and poverty, but also we do not have to rule
out the benefits of genetic engineering, which he asserts can
play an important role in meeting the need for food security.
Put another way, he argues that genetic engineering is needed,
but only if it is developed and introduced as part of a holistic
vision of environmental and socioeconomic sustainability:

Since there is no option in population-rich and land hungry countries but to
produce more per units of land, water and labour, there is a need for technologies which can promote and sustain an ever-green revolution rooted in the
principles of ecology, economics and social and gender equity. It is obvious

---

105 Gordon Conway, *Crop Biotechnology: Benefits, Risks and Ownership*, Paper presented to the OECD Conference on the Scientific & Health Aspects of Genetically Modified Foods, Edinburgh, 28 February to 1 March 2000.
106 As one report refers to Swaminathan: Panos Briefing, 'Greed or need?
Genetically modified crops'. *Panos Media Briefing No 30*, October 1998, p. 8.

that the challenge can be met only by integrating recent advances in molecular genetics and genetic engineering, information and space technologies and ecological wisdom, resulting in appropriate ecotechnologies. There should be no relaxation of yield-enhancing research, since there is no other way of meeting global food needs.[107]

We would be supportive of the benefits of genetic engineering of crops undertaken in this holistic manner, in combination with other conditions and guarantees, which will be spelled out below. However, the fact is that genetic engineering is not being introduced according to the right criteria. In the previous chapter we assessed the current field of play and demonstrated that it is essentially in conflict with the approach advocated by Swaminathan. Those driving forward biotechnology have different priorities, motives and values. As has been indicated, technology is not value-free, and those underlying values can affect whether a new technology is developed and introduced either for the well-being of humanity or not, and alternatively whether it will be for the benefit of an elite or available for everyone. As one African research centre asks, 'can this [biotech] revolution be harnessed to serve the food and nutritional needs of the world's poor?' They believe it can if managed properly, taking into account all the opportunities, problems and risks.[108]

Biotechnology Priorities and the Needs of the Poor

The majority of GM developments to date are primarily focused on research and products beneficial to developed nations. For example, the first wave of products has concentrated on herbicide resistant crops for farmers in order to give

107 M. S. Swaminathan, 'ICRISAT in the 21st Century: Towards Sustainable Food Security', *Environmental Awareness,* Vol. 20, No 4, October–December 1997, Baroda, India.

108 Gabrielle J. Presley & John J. Doyle, *Biotechnology for Developing-Country Agriculture: Problems and Opportunities: Overview,* Focus 2: Brief 1 of 10, October 1999, International Food Policy Research Institute (IFPRI), Washington.

extra life to their particular brands of herbicide like glyphosphate and glufosinate, which farmers must agree to buy. Gordon Conway agrees when he affirms that the focus of the life-science companies has been on developed country markets where potential sales are large, patents are well protected and the risks are lower. Likewise, many point out the fact that most of the current GM crops have been either cash crops grown in developing countries, such as *Bt* cotton in China, or crops that are used for cattle feed such as soybeans and maize. Research and development is also aimed at results that benefit developed nations over poorer ones, such as crops designed to have improved characteristics suitable for either processing, increasing shelf lives of fruits, or resistance to frost and drought.

Research and development on crops beneficial to small farmers tends to be ignored, not because of some global conspiracy, but due to basic economics. It is not good commercial sense to use shareholders' money in ventures guaranteed to produce low returns. Businesses are not run as charities. Some do, however, promote small amounts of research for non-commercial reasons, and on occasions do waive patent rights or royalties for developing country situations. It is not surprising then that both the Church of Scotland study[109] and a *GeneWatch* report concluded that the majority of GM crop developments are mainly being applied to crops of importance to the *developed* world. They fit 'comfortably into modern foods systems that emphasise food processing, consumer niche markets and production efficiency'.[110] By comparison to the need, it is remarkable how little is done that is directly relevant to less-developed countries, and even these are said to be minimally resourced.[111] Finally, they assert that in 1997 'the World

109 D. Bruce, and A. Bruce, (1998) op. cit, pp.177-8.
110 Quoted in *GeneWatch*, Briefing Number 3, 'Genetic Engineering: Can it Feed the World?', August 1998, taken from Rissler, J. and Mellon, M., *The Ecological Risks of Engineered Crops*, MIT Press: Cambridge MA, 1996.
111 J. Puonti-Kaerlas, 'Cassava Biotechnology', *Biotechnology and Genetic Engineering Reviews* 15, pp. 329–364, 1998; 'Genetic Engineering: Can it Feed the World?', *GeneWatch*, Briefing Number 3, August 1998.

Panel on Transgenic Crops concluded that technology transfer projects between multinational corporations and less developed countries were so rare that the examples they cited were '*exceptional*'. At best, therefore, it seems that application to such countries will be largely incidental, arising from so-called 'spillover innovations'.[112] The picture has improved somewhat since then. Some corporations are working in partnership on projects that might genuinely benefit the poor, but it is the exception against the dominant GM focus on markets in the rich countries who are already well fed.

A part of the problem here, some claim, is due to the infancy of the technology, because a lot of the problems in developing countries require more complex work, whereas the work required to develop herbicide tolerance or pest resistance requires modification of say a single trait. It has been suggested that biotechnology applications in agriculture are in their infancy. The first generation of genetically engineered plant varieties have been modified only for a single trait such as herbicide tolerance or pest resistance. The rapid progress being made in genomics will transform plant, tree, and livestock breeding as the functions of more genes are identified. Breeding for complex traits such as drought tolerance, which is controlled by many genes, should then become common. This is an area of great potential benefit for tropical crops, which are often grown in harsh environments and on poor soils.

Research into nitrogen fixing plants from the air would also be invaluable to farmers in poor soil areas, but may prove too difficult technically.[113] However, the question remains, would these developments be pursued to benefit developing countries' farmers, or will they be exploited? This sort of question can only be answered and resolved through dialogue involving

112 H. W. Kendall, R. Beacht, T. Eisner Gould, J. S. Schell, and M. S. Swaminathan, *Bioengineering of Crops: report of the World Bank Panel on Transgenic Crops*, International Bank for Reconstruction and Development/World Bank: Washington DC., 1997.

113 D. Bruce and A. Bruce, (1998) op. cit, pp.26.

all stakeholders, and their analysis and assessment of the problems and potential benefits and risks.

Strategic Conflicts of Corporate Goals and Human Needs

One area of deeply held concern regards the behaviour of corporations in developing countries, where, to further their own interests, they may conflict with what is in the interests of the people involved. The logical commercially-driven question 'What can we do to help that will reward us commercially?' may prove a threat to the well-being and welfare of small farmers. For instance, some argue that if GM varieties fail for some reason, small farmers will be less able to bear the costs of the loss, and could find it increasingly difficult to obtain seeds suited to their needs. The end result could see farmers either falling into debt or being forced out of farming altogether.

It would not be true to suggest that all companies are rapacious or callous, and that worst case scenarios should be the basis on which to judge. Nonetheless, we set out below several technical and strategic developments which seem to us to sow seeds of injustice rather than blessing.

Terminator or traitor technologies or genes ('T&T')

This technology has led to the greatest outcry in developing countries as well as amongst many scientists, citizens, NGOs and campaigning groups, and with good reason. The technology can be described as falling into three phases of development. The first concerns seed sterilisation. The second is about preventing seeds from growing unless counteracted by a chemical spray. The third phase either restricts positive seed attributes or promotes negative attributes, unless counteracted by a chemical spray sold by the company. The complaint is that T&T seems deliberately designed to lock farmers into a cycle of buying seed and chemical spray every season, and this is particularly valuable in developing countries where patent rights or the law are weak. The green NGO RAFI claim that a spokesman for the US Department of Agriculture suggested that the purpose behind T&T technology was 'to increase the value of proprietary seed owned by US seed companies, and to open up new markets in Second and Third World countries'. As a result, he wanted the technology to be 'widely licensed and made expeditiously available to many companies'.[114] This goal was reinforced by a press release of the Delta and Pine Land Company which developed one version of this technology. It claims the terminator technology has 'the prospect of opening up significant world-wide seed markets to the sale of transgenic technology for crops in which seed currently is saved and used in subsequent plantings'. And that 'we expect [the new technology] to have global implications, especially in markets or countries where patent laws are weak or non-existent'.[115]

There would be a major injustice in seeking to apply terminator and related technologies to the situation of the majority of subsistence farmers in developing countries who normally expect to resow their seeds. In the UK, where the normal practice is to purchase new hybrid seeds each season, T&T is not such an important issue. Indeed it may have some potential as a means of preventing unintended gene flow to non-GM species. Yet these technologies have created the greatest outcry and objections from developing countries. As a result, a number of biotech multinationals have begun to deny they are working on T&T technology. RAFI reported that: 'Monsanto's decision to back away from terminator technology, after prompting from Rockefeller Foundations president, Gordon Conway, was an important step. It's not every day that a major multinational enterprise caves in to public opposition

---

114 RAFI Communique, 'The Terminator Technology, New Genetic Technology Aims to Prevent Farmers from Saving Seeds', March/April 1998. Source: http://www.rafi.org.
115 Ibid.

and 'rejects' a new technology. Over the past year, Monsanto, the world's most visible and notorious corporate crusader for genetically engineered seeds, has been battered and bruised by the anti-biotech backlash. Pulling the plug on Terminator was a desperate attempt by Monsanto to distance itself from what is universally considered the most morally offensive application of a biotechnology (so far)…'[116]

In August 2001 the US Department of Agriculture announced that it would be licensing its T & T technology to the Delta and Pine Land Company for development. This is despite the recent view of an FAO panel that terminator seeds are unethical. Such examples do not provide confidence in the industry, and present a strong case for international legislation banning the technology, at least for developing country use.

Aggressive purchasing of local and international seed companies

The buying of seed companies not only gives a company access to seeds and other valuable information. They also gain access to and share of local markets. By way of example, Monsanto, through two purchases, captured 30 per cent of the Brazilian maize seed market and over half in Argentina. Through the acquisition of Cargill's International Seed division it gained seed testing operations in 24 countries and seed multiplication and distribution operations in 51 countries.

Aggressive exploitation of patenting potentially useful genes

Patenting is a negative right which prevents others from commercially exploiting one's own invention. Its extension into genes, plant varieties and transgenic organisms has been highly controversial, and many churches on both sides of the Atlantic have objected to the trend. By gaining access to genetic material in developing countries, making adaptations or in some cases simply finding an industrial use, companies

---

116 Ibid.

have been able to file patents on the genetic resources without any compensation to the originators of the seed or material. This process ensures that anyone wanting to use genes that have been patented must pay for it. This is known as 'biopiracy' and is now outlawed in the UN Biodiversity Convention – but not before many abuses have occurred. The failed attempt to patent Basmati rice was a notable example. Even where patents are legally obtained, it can have a serious detrimental effect on developing country applications in genetics. In the case of Vitamin A rice, the inventor had to use 24 patented technologies, and unless biotech multinationals donate such technologies users will have to pay for them. In this particular case, nearly all the patent rights have been waived, but even one retained patent can keep a potentially beneficial innovation out of the reach of the poor in developing countries.

The current trend indicates that biotechnology companies and Western governments are seeking to strengthen the law on international intellectual property rights (IPRs) in their favour. If countries become members of the World Trade Organisation (WTO), under its 'Trade Related Aspects of Intellectual Property Rights' (TRIPS) chapter they must agree to setting up a system for patenting in their country. Pressure is being applied to ensure that this will include the type of genetic resources of interest to the major biotechnology companies. This strategy will provide a formidable mechanism of control over the use of GM and non-GM plants. Some interpret this as the industry 'positioning itself to dictate the future of plant breeding'.[117] The TRIPS agreement is being renegotiated, however, and there are signs that plenty of developing countries are wise to the situation. A group of African states is arguing that no gene patenting should be allowed in the new agreement, for example. Patenting has a valid place when applied to true 'inventions' (as opposed to discoveries of nature, like genes or

---

117 RAFI Communique, 'Seed Industry Consolidation: Who Owns Whom?', July/August 1998. Source: http://www.rafi.org.

plant varieties), in terms of protecting intellectual property and stimulating innovation. However, the serious downside is that it favours those countries with a strong innovation base, and is strongly geared to large private sector corporate interests. It is currently more likely to disadvantage most of the developing countries of the world than benefit them.

A counter measure to the dominance of corporate interests has been the recent signing of the Biosafety Protocol of the UN Biodiversity Convention. This serves primarily to ensure that countries receiving GM crops have the capacity to assess the potential risks involved, but it also includes significantly greater protection to developing countries in the realm of intellectual property and biopiracy. This was not won easily, as a number of groups representing the interests of the biotech industry and US government tried to weaken the protocol. There is now a deep international dispute over whether the Biosafety Protocol has precedence over any WTO rulings. If it does, it will indeed be beneficial to any nation that wished to stop, or consider longer regarding the benefits of GM technology for their people, without facing a trade war. If not, then the corporations have indeed a strong grip on whether a country is free to dictate its own internal policies and approach to GM technology, crops, and food.

At the end of July 2000 the USA threatened a trade war with Europe if they did not climb down over the issue of labelling GM products. The biotechnology industry argued in law that in the US context to label food as genetically modified was pejorative, implying a warning that something was wrong. It was prejudicial to business interests unless harm could be proved. Because so much American food now contained GM ingredients it was going to destroy their international trade in food products. It was therefore portrayed by the Americans as a form of trade protection, in order to allow indigenous European GM products to gain a share of the market. This was a false argument to apply to a grassroots consumer rejection, in the light of which the EC is now proposing even stronger labelling requirements.

A Way Forward?

If biotechnology is going to be of benefit, the multinationals' level of control needs to be reduced, while allowing them a reasonable level of protection. As a priority, banning the use of T&T technologies in developing countries should be seriously considered, as discussed already. Measures to prevent biopiracy need to be policed and enforced. Countries entering into genetic resource agreements with companies need to safeguard their rights and interests, and insist on benefit sharing.[118] Furthermore, the Biosafety Protocol in practice must have precedence over any TRIPs regulations and rulings. Finally, part of the problem can be solved, Conway maintains, by allowing for a mixed approach:

> On the one hand, they [developing countries] could encourage the for-profit sector to develop and market high quality, locally adapted, premium seeds (especially hybrid seed, which farmers can, if they wish, keep for the next season's crop although the yields are likely to be lower) for the commercial and semi-commercial farmers. Protection would be through a modified PVP (Plant Variety Protection) system. On the other, they would encourage a strong public sector seed system that serves poorer farmers. This would provide an economic incentive for private sector research, innovation, and marketing, and help ensure that the public sector had access to new technologies. Over time, more and more farmers from semi-commercial sectors should be able to buy seeds on a regular basis. And hopefully, many farmers would move from being really poor to the semi-commercial or commercial categories. A key part of such an approach would be the stimulation of public-private partnerships whereby genomic information and technologies are made available to public plant breeders. For example, it is hoped that the patent holders will donate the technologies used in promoting vitamin A rice.[119]

This partnership approach has many merits. Developing countries could then see themselves as stakeholders in the

---

118 S. K. Brahmachari, *Gene patenting: An Indian Perspective*, UNESCO International Symposium on Ethics, Intellectual Property and Genomics, Paris, 30 January – 1 February 2001.
119 Gordon Conway, 'Crop Biotechnology: Benefits, Risks and Ownership', op. cit., OECD 2000

technology's development and in partnership with private-sector companies to genetically modify crops consumed by the poor so that they grow better and more abundantly in their environment without making it worse. Clive James of ISAAA at Cornell University remarks that 'the Nuffield Council on Bioethics concluded that a compelling moral imperative exists to make transgenic crops available to developing countries that want them to combat hunger and poverty. Creative partnerships between developing countries, CGIAR centres, and the private sector could provide the institutional mechanism for sharing the new technologies'.[120]

This has implications for policy making at all levels in order to deliver solutions for the poor that include the benefits of biotechnology. Presley and Doyle in their briefing for the biotechnology industry and WTO propose the following:

> The successful application of modern biotechnology to the problems that cause under-nourishment and poverty could be called a biosolution. The delivery of new biosolutions to the problems of food security and poverty will require continual policy development and actions at the national, regional, and international levels. These efforts will involve the following five areas: (1) determining the priorities and assessing the relative risk and benefits in consultation with the poor, who are often overlooked while others decide what is best for them; (2) setting policies that benefit the poor and minimise technology-transcending risks that adversely affect the poor; (3) establishing an environment that facilitates the safe use of biotechnology through investment, regulation, intellectual property protection, and good governance; (4) actively linking biotechnology and information technology so that new scientific discoveries world-wide can be assessed and applied to the problems of food insecurity and poverty in a timely manner; and (5) determining what investments governments and the international development community will have to make in human and financial resources in order to ensure that biosolutions to the problems of food security reach the poor.[121]

---

120 Clive James, 'Global Status of Commercialized Transgenic Crops: 1999.' *ISAAA Briefs No. 12: Preview.* ISAAA: Ithaca, NY.

121 Gabrielle J. Presley and John J. Doyle, *Biotechnology for Developing-Country Agriculture: Problems and Opportunities: Overview,* Focus 2: Brief 1 of 10, October 1999, International Food Policy Research Institute (IFPRI), Washington.

Perhaps the first priority for GM crops should now be those that benefit the poor and their livelihoods. There may be valid roles for crops suitable for Western demands, such as producing vaccines in plants, environmental improvements in agriculture and possibly nutritional improvements, but these are arguably of second-order importance. If these requirements cannot be met in a developing country, then that society has to decide whether the risks of allowing GM developments can be justified, until a time when it can be confident that biotechnology can be pursued in the recommended way.

9

Reflections

How Should We Evaluate GM Crops and Food?

Genetically modified crops and food has struck such a chord in society because it is an indicator for a set of deeper questions. These concern human intervention in God's creation, the role of science, the right way to do agriculture, risk and precaution, global poverty and world food supply, and the power structures and inequities of the commercial and political world. Christians are divided over these matters, but this study has sought to set out some of the main issues in the light of Scripture, and of Christian insights into science, the environment, agriculture and the needs of developing countries.

Summary of our Findings

We have tried to clarify which questions are specific to GM crops and food, which are common to selective breeding or other areas of industry, and which may be common, but are especially focused by, the GM case. We have examined the science of both modern GM technology and age-old genetic selection in crops, and discovered that many of the concerns expressed about the one could also be raised about the other. Unnaturalness, novelty, safety, gene spread and biodiversity loss are not new issues. The technical ability to switch genes across species that do not normally mate does, however, mark a significant change, whose greater specificity and speed can bring both benefits and risks. There is a general lack of knowledge of what goes on in agriculture, food production and genetic science, which has made the public debate difficult. However, a bigger problem has been the attitude of many in science, industry and government which has dismissed public values as irrational. We critiqued their assumption of the supremacy of scientific rationality as the final arbiter on GM food issues. We pointed to the importance of identifying the underlying values which drive everyone involved, on both sides of the debate, and to bring these into the light of a biblical understanding.

The theology of creation provides a starting point to consider whether genetic engineering is a right or wrong intervention. The notion of 'stewardship' seeks to balance God's calling to exercise dominion with our responsibility of care for the garden of creation. Christians hold differing views about where the balance should lie. We found little substantial biblical evidence, however, to support an intrinsic objection to genetic engineering, either in equating the notion of 'kinds' in Genesis 1 with strict barriers, or in the texts in Leviticus 19 and Deuteronomy 22 about not mixing seeds. The Old Covenant commands emphasise Israel's ritual and genetic separateness from the Canaanite peoples, but are superseded by the finished work of Christ, in the light of which Paul declares all foods to be clean. God's creation is also dynamic rather than static. Scripture shows humans intervening in nature in many ways and we doubt that genetic modification is inherently different, theologically.

We have noted two secular pressures, one to exalt science and the other to retreat from it. One is the drive of scientific determinism, which exalts human intervention and mastery in creation, without Christian values to critique and guide what is done in the name of science. An equal and opposite concern is the rise of neo-pagan views, which elevate nature to a quasi divine status and presume that human intervention is tampering, or the more general presumption that 'nature knows best'.

We caution Christians against either trend. Human intervention is disfigured by the Fall, but God has not revoked the creation ordinances, and nature is not a good guide to morals. Certain aspects of technology have indeed evidenced an excessive intervention, and Christians are encouraged to become involved in environmental issues as an essential part of their calling to serve Christ in our generation. But we should avoid overreacting so much that we go to the opposite extreme and demonise technology. A Christian outlook regards human activity in creation as valid, although tainted by the Fall, in the light of Christ's redemption and resurrection, when guided by the wisdom of the Holy Spirit. This poses some critical questions, however, which we have examined.

We considered whether genetic engineering represents good stewardship in terms of risk. We looked at some examples, but our aim was to examine the principles involved as Christians evaluate risk. We noted that God did not give humanity omniscience, and yet has called us to be creative, which implies risk taking in all spheres of human creativity. We suggest that Christians should not reject GM if there is the least conceivable risk, because no human activity can be risk free. Such a position is not consistent with how God has made the world. We believe God encourages exploration and discovery, but in a context of wisdom and prudent risk management. Precaution offers a wise path, but is nevertheless not a simple answer, because fallen humanity is as able to exaggerate risks that are trivial as to fail to see risks that are serious. Field trials of GM crops can pose some of the very risks they are hoping to test, but on balance we regard them as important to help establish what are, and are not, serious risks. A general moratorium is seen by some as a breathing space to think. Others see it as too indiscriminate because it lumps big and small risks into the same category, and represents an unrealistic counsel of 'perfection', because GM decisions will inevitably have to be made without ever having full data. If, as many Christians believe, there are to be potential benefits from GM crops, delays can also have moral consequences.

The most significant theological critique of genetically modified food arises from questions of justice, power, the needs of the poor and weak, and for agriculture generally. In a UK context, GM crops present both advantages and disadvantages for the farmer, and have potential both to improve or worsen the environment. Against the context of high chemical input conventional farming, organic agriculture has much to commend it as a system which aims to work more harmoniously with nature. It has philosophical elements concerning which we should be cautious, but many Christians have found its practical approach a better response than conventional agriculture. It has its own problems with regard to risk, and its claims to be better for both the environment and health are largely unproven one way or the other. We also question whether it can realistically feed an eventual 8.5 billion people. We caution against viewing it as a simple choice between GM and 'organic', however, and indeed see this way of framing the issue as somewhat politically loaded. The future direction of agriculture has many other methods which seek a more environmentally sustainable approach, and which do not necessarily exclude GM methods, if appropriate. The choice is amongst a spectrum of methods of which organic is only one.

We examined the power structures involved in GM food production and found much to disturb us. The UK GM food episode reveals deeply disturbing questions about who controls genetic developments and for whose ends, if public and private power structures could combine to get things so wrong democratically. It revealed arrogance, unaccountability and undue power on the part of multinational companies and governments alike, who forced their agendas on to an unwilling public. Their collective dismissal of public ethical concerns as irrational proved to be a gross misjudgement. The public rejection has amounted to a subversion of power structures which has reverberated all over the world. There are some signs of change and a willingness to engage with public concerns in the UK, but this may be too late.

The moral claim for needing GM to 'feed the world' has also been deeply compromised by the behaviour of the companies and power structures which promote the technology. Most current applications are for production efficiency in Western supermarket foods, not to feed the truly hungry. GM policies actually manifest the very imbalance of rich and poor which it is supposed to help solve. There seems to be some potential from GM for the poor in developing countries, but only if the right type of application can be made, sensitive to the local needs and the indigenous agricultural, economic and social context. These are big 'ifs', given how far the aims of GM have been dominated and subverted by goals of economic profit and competition. Addressing human and environmental needs, in touch with the views and values of ordinary people, should become the priorities instead. In this context, we advocate that, as Christians, we should earnestly examine our own involvement with consumerism and its values.

How shall we Evaluate GM Crops and Food?

GM crops and foods present a complex ethical case to assess. It resembles a diamond with many facets, depending on which angle we examine it from. Can we stand back and evaluate it as a whole? We have presented our findings by way of instruction and explanation, offering our insights from our differing experience and expertise. On many of these issues we differ among ourselves, and so it would be inappropriate to present 'our definitive conclusion' on GM crops and foods. We invite the reader to make up his or her own mind.

One way of posing the question is to ask first if there is an absolute case for rejecting GM as wrong in its very act. If not, then the issues become largely ones of consequences, and we must decide what are the criteria which stand out as crucial? Some, like Christian Aid, maintain that corporate injustice and abuse are so endemic to the whole GM venture to be cause enough to reject it, whatever its benefits might have

been.[122] For some environmentalists, GM represents a technical 'fix', which perpetuates the mistaken trend of intensive
industrial agriculture and deflects the focus from more
important questions about sustainable agriculture. Many
Christians may find common cause with these various NGO
conclusions, and regard the political-economic context as an
overwhelming case against GM food, because its corporate
'package' is too flawed to be morally acceptable.

Some would argue that GM crops, like technology in general, are 'here to stay' and a part of contemporary society. Even
if they do not approve, Christians should 'work within the system', to control and constrain, to challenge and change, to seek
to bring the development and application of GM technology
into an ethical and environmentally responsible framework.

Others, however, regard the key criteria as the benefits to the
poor and the environment. Can a redemptive aspect to GM
food technology still be found? In particular, they would see
developments like GM vitamin A rice as representing the type
of application which has potential to make a real difference to
nutrition in the developing world, notwithstanding all the
structural issues which need to be addressed concerning poverty. Ultimately we in the well-fed West cannot prohibit other
countries from employing our technology if they consider it
might answer some of their problems. As Christians, should we
not simply be satisfied with the prophetic role of condemning
corporate dominance of the direction and control of GM
technology, and carry out the redemptive task of looking for
viable alternative ways of prioritising, delivering and managing
it? In the UK context, whilst organic methods may represent a
romantic reaction to one problem, do they involve their own
potential risks which have not been examined as critically as
those of GM crops? Can GM technology play an important
role in environmentally sustainable agriculture that can draw
from the best of both worlds? Such questions lead some
churches, which express caution about GM risks and revulsion

---

122 Christian Aid, *Selling Suicide.* op. cit., 1999.

at injustices in how GM has been applied, to declare nonetheless that we should not 'throw the baby out with the bathwater.'

Another possible approach may arguably be derived from a study of Old Testament ethics. Chris Wright has suggested three possible responses of the people of God to aspects of culture and society.[123]

- *Accept and affirm.* Some aspects of culture can be 'taken over' or 'reclaimed' by God's people and be positively affirmed. Wright cites the arts, and the 'disinfecting' of pagan festivals like Christmas and Easter, and most important, the family.
- *Tolerate, but work for change.* Wright cites polygamy, divorce and slavery as aspects that were tolerated, but strictly controlled and constrained. Jesus was, 'without compromising his own holiness, able to tolerate the institutions and structures of his day, while at the same time challenging and undermining them with the revolutionary message of God's dynamic rule among men'.[124]
- *Renounce and separate.* 'Some aspects of fallen human society must be rejected as abhorrent to God… the categories of what was socially prohibited to Israel included … in particular: the idolatrous, the perverted, that which was destructive of persons and callousness to the poor'.[125]

Wright suggests that many aspects of secular culture could be placed in these categories, and calls for prophetic denunciation by the Church.

If this framework were followed, there would be a case for placing GM crops and food in any of the three categories, however. Some would see it as falling into category one. If we 'accept and affirm' the creative arts with all their ambiguities,

123 Wright, C. J. H., (1983), *Living as the people of God,* Leicester: Inter-Varsity Press, pp.174ff.
124 *Living as the people of God,* pp.189–90.
125 *Living as the people of God,* p.187.

then the 'practical arts', namely science and technology must similarly be accepted and affirmed, whilst seeking to correct their faults. To make a distinction would be quite unfounded. Others would argue that for farmers, and perhaps society as a whole, the uncertainties as to the balance and distribution of costs and benefits, the environmental risks and the concentration of power and control associated with GM technology could not be affirmed and accepted, and so would place GM crops into category two. Some again would contend that the mixing of genes is inherently wrong, or that the risks were intolerable, or that the power of big business had reached such proportions, that the only conclusion was separation.

There remains some doubt, however, regarding how far we can validly appropriate Wright's framework. Its primary analogy is derived from the theocratic society of the Old Testament, surrounded as it was by alien nations. This does not easily recontextualise into the largely secularised society in which Christians are obligated to work for the kingdom of God. The way it poses the question already tends to presume confrontation between human activity and God's will, rather than starting from God's common grace to all humanity in creation. In the light of the Fall, all human activity is more or less 'spoiled'. One cannot single out GM crops as 'less than the best', because this is true of all technology. It might be possible to distinguish one application of GM crops, or one painting, as less in keeping with God's will than another, but we can no more generalise about one whole area of technology than we can with one whole art form. So this framework, whilst helpful for some, still leaves many of the key questions about GM unanswered.

Some Christians regard the essential issue relating to GM crops and food as resting not so much on what they are, or what they may do, but on what they symbolise. They might identify an analogy with Daniel and his companions in 'refusing the king's food' (Dan. 1:8-16), who did not refuse Nebuchadnezzar's educational programme, nor the changes in name, but baulked over the matter of the king's food. There were several reasons for this, perhaps most significantly, that it would be ritually unclean. Some have suggested they refused the king's food, however, just because it was the king's. Whilst not necessarily accepting and affirming everything in Babylonian society, they were prepared to tolerate much, to work within the system and to respect the king. But they were unwilling to go so far as to have 'table fellowship' with him.[126] For them, perhaps, eating the king's food would somehow represent an acceptance of all that the king stood for, and that represented a step too far. By analogy, GM crops and food symbolise the products of an alien culture, from the world of agribusiness, impacting unwelcomed on a beleaguered farming culture and unwitting consumers. To them, GM appears not so much as representing good gifts from a heavenly Father who loves us and provides us with food and rain, but rather symbolises greed and exploitation, and a wasting of creation. Accordingly resistance is considered justified.

There will be divergent views regarding the validity of such biblical exegesis and interpretation. It may be viewed as wholly inappropriate to highlight the problems of GM as a paradigm for contemporary features of the culture. Its specific applicability to GM may be seen as not only irrelevant but unfair, whilst we remain uncritical of similar issues that we accept unquestioningly in conventional food production, and indeed our culture in general. Christians should aim at practical consistency in what they approve and disapprove. If we reject GM because of its power structures, then arguably we might find ourselves having to disavow large swathes of our culture.

Postscript

For most people GM food is still novel and unfamiliar. As with nuclear power, the notion of genetic engineering can carry sinister connotations. The BSE crisis left a legacy of distrust in

126 R. Clements, *Practising Faith in a Pagan World,* Leicester: Inter-Varsity Press, 1997, pp.103–18.

the regulatory system, and an association of scientific intervention in food with potential danger. This raises a final question. Who should we as Christians believe amongst the clamour of voices we hear? Most Christians derive their knowledge from the news media, secular or Christian, which may or may not be a good guide. The tendency to sensationalise the issues, either way, poses a real concern given how deeply politicised the GM issue has become. We would distinguish our work from a 'campaigning' approach which can 'hold the truth hostage' by focusing only on the aspects convenient to particular political objectives and agendas. Christians have genuine deep-rooted concerns both ways, and passions can run high. We have tried to take serious and respectful account of the differing perspectives, as fairly as possible. We seek to present what we see as a truthful and biblical account of a highly controversial set of issues, informed by our specialist knowledge, but without favouring any particular agenda. In a post-modern context, where 'ethics' is frequently reduced to power structures and the pleadings of special interest groups, Christians have an important role to play. By setting these conflicting single issues into the wider perspective of God's principles for all humanity, Christians surely retain an important apologetic opportunity within our culture.

10

Resolutions and Affirmations of the Evangelical Alliance Policy Commission relating to GM Crops and Food

Genetic modification of crops and food is a highly controversial and emotive issue. Christians differ widely in their responses to the issues it raises. This statement by the Policy Commission seeks to comment responsibly on the debate in the context of a Christian understanding of Creation, based on the report of the working group. We believe that a number of key conclusions can be drawn, and recommendations made, which will win wide support among evangelicals, in spite of their diverse approach on some aspects of this issue, and which we believe could find a much wider endorsement.

The statement should also be read in the context of (and with acknowledgements to) the *Evangelical Declaration on the Care of Creation* (1994) which the Evangelical Alliance has previously endorsed. The *Declaration* is reproduced in its entirety in Appendix 2.

Conclusions and Recommendations

1   The issues raised by genetic modification involve spiritual and ethical, as well as technological, considerations. The

potential technological benefits are important, but as Christians, we believe that the task of care for the earth and its people raises questions which are, above all, spiritual in character. The answers must be sought in God's revelation in Christ and the Scriptures, especially in his purposes for the natural world both in creation and redemption.

2   We live in a society with a predominantly materialistic worldview. We also note the emergence of views reacting against this, especially in the environmental field. We reaffirm the traditional Christian understanding that the world around us is the work of a Creator God, who calls us to act according to the divine principles which he has set for the creation, and to follow wise and responsible public policies which embody biblical principles for the stewardship of creation. This requires realistic and responsible evaluation of human objectives and motivation. It bears on individual lifestyles and behaviour, on the policies and actions of private bodies such as companies, foundations and non-governmental organisations, and on the policies and actions of government authorities. It concerns both consumers and producers. Christians believe that our aim in all cases should be to ensure that we comply with God's ordinances for creation. We ought therefore to affirm and give effect to a practical discipleship which resists wastefulness and over-abundance of possessions. This implies making individual choices about our lifestyle which emphasise humility, self-restraint, and generosity towards those in need.

3   The principle of godly, just, and sustainable economies which enable all people to flourish along with the diversity of the rest of creation, is a fundamental Christian objective. We recognise that human poverty is both a cause and a consequence of environmental degradation, and call upon governments and Christians to give priority to and support measures which protect and improve the

environment, and which use its productive capacity in ways which assist rather than exploit the poor.

4 Regarding GM crops and foods, we recognise that there are differing viewpoints among Christians which lead them to advocate different policies and practices. There should be freedom for such differing viewpoints to be expressed and weighed. We do not, however, support action to prevent legitimately approved trials designed to clarify the uncertainties about the effects of GM crops in the environment. In particular, we oppose illegal action directed against producers and users of GM crops and foods, such as the destruction of trial crops, attacks on supermarkets, and harassment of farmers and their families. We plead for fair-minded truth-telling in campaigning and the dissemination of information, together with competent and open-minded investigation of scientific data, risks, and benefits, and an acknowledgement of, and sensitivity towards, different value positions.

5 We acknowledge that there is common ground at a practical level between the environmental movement and the Christian approach to the care of God's creation. We would encourage Christians to be involved in environmental activity with secular groups in their locality. We differ, however, from the ideological and quasi-religious aspects of certain 'green' agendas where these draw from pantheist or New Age belief systems. Creation care is God-centred, not earth-centred. The Holy Spirit's work in creation is of a very different order to false concepts which seek to idolise 'Nature' and its 'forces', 'energies' and 'harmonies'.

6 In a society in which perceptions of risk are increasingly important, the Evangelical Alliance recommends a cautious, though not anti-scientific, approach. It is important to maintain continued vigilance against any short-cutting of appropriate risk assessment involving the full-scale introduction of GM crops and foods. We consider that the potential benefits of GM technology are too great to

neglect, and fall within the permitted range of acceptable human discovery and experiment. However, we remain concerned that there should be robust national and international ethical frameworks to regulate and limit acceptable developments, and open dialogue and constructive engagement with interested parties. We welcome recent developments in national and international institutions in this field, especially the setting up of the UK Agriculture and Environmental Biotechnology Commission (AEBC) to look at wider ethical issues in a consultative manner. Regulatory performance should be monitored against the principles just enunciated. At an international level, we believe that the rules of the World Trade Organisation need to be amended so that countries can, if they so desire, refuse the importation of GM crops and foods if this conflicts with national ethical norms, safety or environmental evaluations, without fearing trade sanctions. We regret the behaviour of some large corporations and governments which have marginalised or disregarded public anxieties and informed opinion which scrutinises, challenges, and offers different perspectives and values. In particular, we question the 'scientistic' and 'economistic' approaches which often underlie the policies of government, companies and international institutions on this matter.

7   We welcome scientific developments, such as genetic engineering, in so far as they have the potential, when managed and applied wisely, to achieve benefits such as improved and increased food production and the relief of malnutrition. While recognising the underlying structural causes of hunger and poverty, we believe that genetic engineering might also play an important role in helping to achieve food security. But we recognise that the same technology may also be employed recklessly and unwisely for commercial exploitation, and to meet the demands of Western consumerism at the expense of the environment and the needs of developing countries. We therefore oppose deregulation and commercialisation before there has

been proper research and debate. International institutions, governments, and people at the grass roots need to
work towards policies which regulate more closely the activities of market-driven multinational companies in the
field of genetic modification.

8   Given the many uncertainties concerning the environmental impact of GM crops, the regulatory authorities
need to exercise caution in balancing the importance of
farm-scale crop trials to evaluate the impacts of a particular crop, against the potential detrimental effects.

9   We support the development and introduction of geneti-
cally-engineered crops only in the context of a holistic vision of environmental and socioeconomic responsibility.
We consider there is evident need for a moral basis for
technologies which can promote and sustain the global
production of food, rooted in the principles of sound
ecology, economics and social justice. If this challenge is to
be met, it would require much care in applying advances
in genetic engineering in the context of traditional agricultural practices, using applications that benefit the poor
and give farmers a sustainable livelihood. We oppose activities which would result in the benefits of the new
technology becoming unaffordable in developing countries, such as the patenting of gene sequences and transgenic species, and the use of terminator technology to
restrict the re-use of seed in countries where this is the
normal practice. We note that the terminator technology
may in some circumstances have environmental benefits
in a UK context. We understand that companies need to
recover the costs of research and development, but we
urge them to achieve this in morally-responsible ways. In
the absence of such corporate responsibility, international
regulation of their activities may become necessary.

10 Clear freedom of choice for consumers should apply in
respect of GM foods. The principle and practice must
therefore be established of visible, accurate, and informa-
tive labelling of all foods which have involved a process of

genetic modification. To this end we would encourage the Government to continue negotiations with the European Commission to bring in firmer legislation, closing some of the present loopholes in European Union regulations.

11   Although we recognise the arguments and sincerity of many who choose to respond to the implications of the GM issue by preferring organic foods, we nevertheless recommend appropriate discernment with regard to ideologies associated with organic farming.

12   Christians and governments, especially in the West, should seek to ensure that technologies are developed and applied in a way that gives priority to the poor and to developing countries in preference to the interests of the wealthy, in addition to meeting environmental needs. We commend investigation of potential benefit-sharing partnership arrangements with developing countries in which they may act as stakeholders in the technology's development to protect their rights and interests. Complementary workable measures to prevent biopiracy and to safeguard legitimate commercial interests would need to be implemented.

Appendix 1

Glossary of Technical and Unfamiliar Terms

ACRE Advisory Committee on Releases to the Environment – The UK Government regulatory body within the Department of Environment, Food and Rural Affairs, which has statutory responsibility to refuse or to grant the releases of genetically modified crops into the environment.

Agrobacterium A soil bacterium that infects plants and transfers some of its DNA to the host plant.

Allele One of two or more alternative forms of the same gene. In those many organisms (including humans) whose body cells have two sets of genes, it is possible for an individual to possess two different alleles of any particular gene. Also see *polymorphism*.

Backcross The crossing of a new hybrid variety or inter-species hybrid with one or other of the parental varieties or species (see also *hybridisation*, below).

Bt corn/maize Crops in which insect resistance has been induced using genes which produce a natural insecticide from a soil-dwelling bacterium called Bacillus thuringiensis (Bt). Crystals of the same toxin have been used since the 1950s as an organically-approved pesticide sprayed or dusted onto plants.

Cash Crops Crops grown to sell rather than to be eaten by the grower. In developing countries cash crops are not primarily for indigenous consumption, but for export to rich nations, in order to earn foreign currency and improve the balance of payments. Generally refers to crops other than cereals sold without processing.

Chromosome A long string of DNA (i.e. a long DNA molecule) that contains many genes along its length.

DNA The chemical substance of which genes are made; the order of the four different building blocks ('bases') that make up DNA acts as a biochemical code to specify the biochemical make-up and activity of each cell.

Enzyme A biological catalyst that is able to bring about a specific biochemical reaction.

Gene A specific tract of DNA within the length of a chromosome that is 'read' to give a specific biochemical 'instruction', for example, 'make insulin.'

Genome Collective term for all the genes of an organism. Can be applied to species, e.g., the human genome project, or to individuals, e.g., inserting a new gene into the genome of a rice plant.

Green Revolution The name given for a widespread publicly funded initiative beginning in the 1960s which sought to improve yields for farmers in developing countries, using research based technologies and especially high yielding hybrid crop varieties. Its success or failure has long been the subject of much claim and counter-claim.

Hybridisation The crossing of two varieties of a species or even the crossing of two species to make a new variety or a new inter-species hybrid (see also *backcross,* above).

Insulin A mammalian hormone that is involved in the control of sugar metabolism.

Landrace A traditionally used local variety of a particular crop species.

Mutation A change within a gene in the order of the building blocks of DNA.

Nucleus (plural, nuclei) In all living organisms above the level of bacteria, cells are physically divided into compartments with different functions. The chromosomes are located in (and hence the genes are 'read' in) the nucleus.

Pathogen Mainly used in connection with the causal agents of plant disease. It is an infectious agent capable of causing disease. It multiplies in the host (in contrast with, e.g., mineral deficiencies, which are known as 'disorders'). A disease is a harmful deviation from normal physiological functioning, caused by an infectious agent. The agent is an organism or substance which causes disease – typically harmful types of bacteria and viruses. E.Coli 0157 and the active agent in BSE, are examples of pathogens.

Plant variety rights/plant breeders' rights A system of ratification and licensing of new crop varieties providing a royalty to the breeder but without restriction on onward use by the grower.

Plasmid A small 'extra' chromosome, usually consisting of a *circular* DNA molecule, that many bacteria possess.

Polymorphism The existence of two or more forms (*alleles*: see above) of a gene within a population.

Position effect The variation in the level of expression of a gene (i.e. how strongly it works) according to its particular position within the genome (*genome*: see above).

Promoter A tract of DNA adjacent to a gene that is involved in controlling whether the gene is working or not; effectively the gene's 'on-off switch'.

Recombinant DNA A DNA molecule constructed in the laboratory, usually by 'cutting and pasting' genes and promoters from different sources.

TIBRE Targeted Input for a Better Rural Environment – A scheme set up by Scottish Natural Heritage for an environmental approach to farming, based on new technology by promoting more efficient use of chemicals and other inputs, aiming at both commercial and environmental benefits.

An Evangelical Declaration on the Care of Creation

The earth is the LORD's, and the fullness thereof.   (Psalm 24:1)

As followers of Jesus Christ, committed to the full authority of the Scriptures, and aware of the ways we have degraded creation, we believe that biblical faith is essential to the solution of our ecological problems.

- Because we worship and honour the Creator, we seek to cherish and care for the creation.
- Because we have sinned, we have failed in our stewardship of creation. Therefore we repent of the way we have polluted, distorted, or destroyed so much of the Creator's work.
- Because, in Christ, God has healed our alienation from God and extended to us the first fruits of the reconciliation of all things, we commit ourselves to working in the power of the Holy Spirit to share the Good News of Christ in word and deed, to work for the reconciliation of all people in Christ, and to extend Christ's healing to his suffering creation.
- Because we await the time when even the groaning creation will be restored to wholeness, we commit ourselves to work vigorously to protect and heal that creation for the honour and glory of the Creator – whom we know dimly through creation, but meet fully through Scripture and in Christ.

We and our children face a growing crisis in the health of the creation in which we are embedded, and through which, by God's grace, we are sustained. Yet we continue to degrade that creation.

- These degradations of creation can be summed up as: 1) land degradation; 2) deforestation; 3) species extinction; 4) water degradation; 5) global toxification; 6) the alteration of atmosphere; 7) human and cultural degradation.
- Many of these degradations are signs that we are pressing against the finite limits God has set for creation. With continued population growth, these degradations will become more severe. Our responsibility is not only to bear and nurture children, but to nurture their home on earth. We respect the institution of marriage as the way God has given to insure thoughtful procreation of children and their nurture to the glory of God.
- We recognise that human poverty is both a cause and a consequence of environmental degradation.

Many concerned people, convinced that environmental problems are more spiritual than technological, are exploring the world's ideologies and religions in search of non-Christian spiritual resources for the healing of the earth. As followers of Jesus Christ, we believe that the Bible calls us to respond in four ways:

- First, God calls us to confess and repent of attitudes which devalue creation, and which twist or ignore biblical revelation to support our misuse of it. Forgetting that "the earth is the Lord's," we have often simply used creation and forgotten our responsibility to care for it.
- Second, our actions and attitudes toward the earth need to proceed from the centre of our faith, and be rooted in the fullness of God's revelation in Christ and the Scriptures. We resist both ideologies that would presume the Gospel has nothing to do with the care of non-human

creation and also ideologies that would reduce the Gospel to nothing more than the care of that creation.

- Third, we seek carefully to learn all that the Bible tells us about the Creator, creation, and the human task. In our life and words we declare that full good news for all creation which is still waiting "with eager longing for the revealing of the children of God" (Rom. 8:19).
- Fourth, we seek to understand what creation reveals about God's divinity, sustaining presence, and everlasting power, and what creation teaches us of its God-given order and the principles by which it works.

Thus we call on all those who are committed to the truth of the Gospel of Jesus Christ to affirm the following principles of biblical faith, and to seek ways of living out these principles in our personal lives, our churches, and society.

- The cosmos, in all its beauty, wildness, and life-giving bounty, is the work of our personal and loving Creator.
- Our creating God is prior to and other than creation, yet intimately involved with it, upholding each thing in its freedom, and all things in relationships of intricate complexity. God is *transcendent*, while lovingly sustaining each creature; and *immanent*, while wholly other than creation and not to be confused with it.
- God the Creator is relational in very nature, revealed as three persons in One. Likewise, the creation which God intended is a symphony of individual creatures in harmonious relationship.
- The Creator's concern is for all creatures. God declares all creation "good" (Gen. 1:31); promises care in a covenant with all creatures (Gen. 9:9-17); delights in creatures which have no human apparent usefulness (Job 39-41); and wills, in Christ, "to reconcile all things to himself" (Col. 1:20).
- Men, women, and children, have a unique responsibility to the Creator; at the same time we are creatures, shaped

by the same processes and embedded in the same systems of physical, chemical, and biological interconnections which sustain other creatures.

- Men, women, and children, created in God's image, also have a unique responsibility for creation. Our actions should both sustain creation's fruitfulness and preserve creation's powerful testimony to its Creator.

- Our God-given, stewardly talents have often been warped from their intended purpose: that we know, name, keep and delight in God's creatures; that we nourish civilization in love, creativity and obedience to God; and that we offer creation and civilization back in praise to the Creator. We have ignored our creaturely limits and have used the earth with greed, rather than care.

- The earthly result of human sin has been a perverted stewardship, a patchwork of garden and wasteland in which the waste is increasing. "There is no faithfulness, no love, no acknowledgement of God in the land ... Because of this the land mourns, and all who live in it waste away" (Hos. 4:1,3). Thus, one consequence of our misuse of the earth is an unjust denial of God's created bounty to other human beings, both now and in the future.

- God's purpose in Christ is to heal and bring to wholeness not only persons but the entire created order. "For God was pleased to have all his fullness dwell in him, and through him to reconcile to himself all things, whether things on earth or things in heaven, by making peace through his blood shed on the cross" (Col. 1:19-20).

- In Jesus Christ, believers are forgiven, transformed and brought into God's kingdom. "If anyone is in Christ, there is a new creation" (2 Cor. 5:17). The presence of the kingdom of God is marked not only by renewed fellowship with God, but also by renewed harmony and justice between people, and by renewed harmony and justice between people and the rest of the created world. "You will go out in joy and be led forth in peace; the mountains and the hills will burst into song before you, and all the trees of the field will clap their hands" (Isa. 55:12).

We believe that in Christ there is hope, not only for men, women and children, but also for the rest of creation which is suffering from the consequences of human sin.

- Therefore we call upon all Christians to reaffirm that all creation is God's; that God created it good; and that God is renewing it in Christ.
- We encourage deeper reflection on the substantial biblical and theological teaching which speaks of God's work of redemption in terms of the renewal and completion of God's purpose in creation.
- We seek a deeper reflection on the wonders of God's creation and the principles by which creation works. We also urge a careful consideration of how our corporate and individual actions respect and comply with God's ordinances for creation.
- We encourage Christians to incorporate the extravagant creativity of God into their lives by increasing the nurturing role of beauty and the arts in their personal, ecclesiastical, and social patterns.
- We urge individual Christians and churches to be centres of creation's care and renewal, both delighting in creation as God's gift, and enjoying it as God's provision, in ways which sustain and heal the damaged fabric of the creation which God has entrusted to us.
- We recall Jesus' words that our lives do not consist in the abundance of our possessions, and therefore we urge followers of Jesus to resist the allure of wastefulness and over-consumption by making personal lifestyle choices that express humility, forbearance, self restraint and frugality.
- We call on all Christians to work for godly, just, and sustainable economies which reflect God's sovereign

economy and enable men, women and children to flourish along with all the diversity of creation. We recognise that poverty forces people to degrade creation in order to survive; therefore we support the development of just, free economies which empower the poor and create abundance without diminishing creation's bounty.

- We commit ourselves to work for responsible public policies which embody the principles of biblical stewardship of creation.
- We invite Christians – individuals, congregations and organisations – to join with us in this evangelical declaration on the environment, becoming a covenant people in an ever-widening circle of biblical care for creation.
- We call upon Christians to listen to and work with all those who are concerned about the healing of creation, with an eagerness both to learn from them and also to share with them our conviction that the God whom all people sense in creation (Acts 17:27) is known fully only in the Word made flesh in Christ the living God who made and sustains all things.
- We make this declaration knowing that until Christ returns to reconcile all things, we are called to be faithful stewards of God's good garden, our earthly home.

Appendix 3

The UK Regulatory System for GM Crops

The release and marketing of genetically modified crops is governed in the UK by the European directive 90/220, a revised version of which has recently been agreed and will be implemented within the next 18 months. The use of genetically modified materials in foods is governed by the EC regulation 258/97 on novel foods and novel food ingredients and by the UK regulations detailing how the European regulation is to be applied in the UK. These regulations have the force of law.

The central purpose of this EU legislation is to ensure that genetically modified organisms should not cause harm to the environment or to the consumer. Organisms cannot be released into the environment without a consent being issued following a careful scientific assessment. In the UK this is carried out by the Advisory Committee on Releases into the Environment (ACRE), which advises the Minister responsible for the environment. Similarly no foods containing genetically modified ingredients can be marketed until they have been approved at European level also following a detailed scientific assessment. In the UK this is the responsibility of the Advisory Committee on Novel Foods and Processes (ACNFP), which is a committee advising the Food Standards Agency.

Both committees are composed of experts with an independent chairman, normally a senior academic. Although appointed by government, the committee does not have any civil servants as members, although they play an important role in preparing the documents, and in taking forward the committees' advice. Each member of the committee is an expert in one or more of a number of different areas, appropriate for the work of the committee. The ACNFP has a consumer person and an ethical adviser as additional members.

There is an increasing trend for openness and transparency in the workings of these committees. An Annual Report and agendas for the committee meetings have been published for some years. Minutes are now published quickly on the Web, and some meetings are being held in public. The ACNFP now routinely publishes on its website complete applications submitted to it. Members of the public then have the opportunity to comment on the application before the ACNFP has assessed it. The only data not available are those that are regarded as commercially confidential. This represents a difficult area. It is a moot point where the balance should lie between public and commercial confidence.

Rules about conflict of interest have always been strict, and are increasingly carefully examined under the Nolan report requirements. Any interest must be declared before discussion of the item. This is recorded in the minutes, and the member then leaves the room for that item. Once a year, in the Annual Report, all members' interests are recorded, while the chairman is not permitted to have any such interests at all in any area that is relevant to the work of the committee. Committee members receive travel expenses and a very modest daily allowance, but are otherwise unpaid for their work for the committees.

These committees work on a case-by-case basis, receiving applications from companies who wish to market a product in the UK, whether it is seed for a crop, or a food, or food ingredient. Decision trees have been worked out to enable the applicants to know how much information is going to be needed for assessment. It is common for the committees to go back to the applicant to ask for further information, or more experiments. Very few applications are approved at the first sight. The committee's advice on all such applications is then peer reviewed by equivalent committees in the other 14 EU Member States.

Over the last two years, the process has been changed and improved by the appointment of a new body commission, the Agriculture and Environment Biotechnology Commission (AEBC). AEBC was set up to meet a long felt need for an over-view body on the wider ethical and social issues raised by animal and plant biotechnology and genetic engineering. It includes a broad range of disciplines – scientific, ethical, sociological, consumer and environmental – and also differing viewpoints, including both those favouring and opposed to GM technology. AEBC oversees the work of ACRE but is involved in case-by-case assessment of individual applications for new products or processes. It is intended to give advice to Ministers on the wider picture and act as a focus for public debate in respect of biotechnology. Regular consultation with stakeholders and the public is seen as an essential aspect of its work.

These changes had been made because it became clear that the advisory committees, as they were constituted, did not carry the confidence of the British public. The new processes are intended to be open, transparent and inclusive, seeking to re-establish confidence in the regulatory system.

Appendix 4

Old Testament Law on Seeds

Summary: Taking Leviticus 19:19 and Deuteronomy 22:9-11 together, it appears that the overwhelming purpose of these laws, both in terms of their content and their context, is cultic and they are designed to maintain Israel's distinctiveness vis à vis the rest of the nations. This meant maintaining cultic and sexual purity and refraining from marriages with foreign peoples. It does not exclude the possibility that the laws also reflect the need to preserve natural distinctions created by God; however this motivation is nowhere explicit within the text.

Laws of Leviticus

Leviticus 19:19

> You shall keep my statutes. You shall not let your cattle (Heb. *behemot*) breed with a different kind (*kileayim*); you shall not sow your field with two kinds (*kileayim*) of seed; nor shall there come upon you a garment of cloth made of two kinds (*kileayim*) of stuff.

No motivating clause attaches to these laws: they are to be obeyed because YHWH commands it (Lev. 19:19a) and no further explanation is given. The prohibition is against mixing 'two kinds', as in Leviticus 19:19c and Leviticus 19:19d. The same phrase – 'two kinds' (in Hebrew *kileayim*) is used in all three prohibitions, indicating that the three prohibitions are interlinked. This is important and signifies that the theme is the general one of 'mismating'.

Notably, Leviticus 19:19 is followed by Leviticus 19:20 which is a warning against an illicit human sexual relationship. In other words, Leviticus 19:19 extends wider than the agricultural and cultural practices of the day to encompass *human* inter-breeding and sexuality.

Leviticus 19:19b

The word translated 'cattle' is the generic word for a 'beast' in Hebrew (*behemah*). Doubtless, the typical prohibition in Leviticus 19:19b is against mating domestic animals and raising hybrid creatures such as mules. The sexual nature of the prohibition (Lev. 19:19b uses the verb for 'copulation') anticipates the law regarding sexual relations in Leviticus 19:20-22.

This raises the question whether the laws in Leviticus 19:19 are simply restricted to agricultural (Lev. 19:19b and Lev. 19:19c) and cultural practices (Lev. 19:19c), or whether they are also meant to teach us something about human sexuality more generally. Following Paul in 1 Corinthians 9:9-10, we might say that Leviticus 19:19b is not simply about oxen, but also has meaning for human relationships:

> For it is written in the Law of Moses, 'You shall not muzzle an ox when it is treading out the grain.' Is it for oxen that God is concerned? Does he not speak entirely for our sake? It was written for our sake, because the plowman should plow in hope and the thresher thresh in hope of a share in the crop.

Accordingly, the warning in Leviticus 19:19b about 'mismating' cattle may be a warning against 'mismating' with the wrong human partners. Certainly the following verse, Leviticus 19:20, is a warning against 'mismating' with 'a slave girl promised to another man'. The slave girl is probably a foreigner (though this is not the key issue, as clearly it is acceptable for a another Israelite to marry her). Even so, it is possible to read this as a warning against marrying foreigners who do not share in the faith of Israel.

This is certainly consistent with the theology of Leviticus as a whole. The issue is not simply 'mating' different kinds of cattle but the folly of mating believers with unbelievers. If so, it is similar to Paul's warning in 2 Corinthians 6:14-17:

> Do not be mismated with unbelievers. For what partnership have righteousness and iniquity? Or what fellowship has light with darkness? What accord has Christ with Belial? Or what has a believer in common with an unbeliever? What agreement has the temple of God with idols? For we are the temple of the living God; as God said, 'I will live in them and move among them, and I will be their God, and they shall be my people.' 'Therefore come out from them, and be separate from them, says the Lord, and touch nothing unclean...'

Leviticus 19:19c

The following command prohibits sowing two different kinds of seed in the same field. The issue is not merely the *co-existence* of two different 'kinds' in the same field, but their 'mating' as in Leviticus 19:19b. It is a prohibition against chance cross-fertilisation. Two different kinds of seed in a single field could give rise to hybrid forms in the vegetable kingdom. The law thus follows naturally on from Leviticus 19:19b; no mismating in the animal kingdom is followed by no mismating in the plant kingdom.

It is possible that this law reflects agricultural wisdom along the lines, for example, of Isaiah 28:23-29:

> Give ear, and hear my voice; hearken, and hear my speech. Does he who plows for sowing plow continually? Does he continually open and harrow his ground? When he has leveled its surface, does he not scatter dill, sow cummin, and put in wheat in rows and barley in its proper place, and spelt as the border? For he is instructed aright; his God teaches him. Dill is not threshed with a threshing sledge, nor is a cart wheel rolled over cummin; but dill is beaten out with a stick, and cummin with a rod. Does one crush bread grain? No, he does not thresh it for ever; when he drives his cart wheel over it with his horses, he does not crush it. This also comes from the LORD of hosts; he is wonderful in counsel, and excellent in wisdom.

However, in view of the context (the law on 'mismating' in Lev. 19:20) it is likely that Leviticus 19:19c has, in addition, a didactic purpose. It is not merely about plants 'mismating' but about people 'mismating', especially outside the community of faith.

Leviticus 19:19d

On the face of it, this law appears to contradict the instructions given in Exodus 28 regarding the manufacture of the priestly garments. These are an admixture of yarn and linen (Ex. 28:4-5). There may be a distinction between priestly and lay; i.e. what is acceptable for the priests is unacceptable for the laity. Certainly, this distinction is found throughout Leviticus (e.g. the priests are allowed to officiate in the Sanctuary and to consume certain offerings whereas the laity are not).

Yet the fact that mixing materials is not an absolute prohibition binding upon priests as well as people is a further indication that this law, like Leviticus 19:19b and Leviticus 19:19c, is didactic in its purpose. The issue is not simply 'mismating' but 'mismatching'. Its educational purpose may be a symbolic reminder of reverence for the order of nature as determined by God. But in view of what follows, in Leviticus 19:20, and the continuity of expression with Leviticus 19:19b and Leviticus 19:19c, it is more likely that the law is intended to teach against human 'mismatches'; specifically between the people of God and pagan foreigners.

Laws of Deuteronomy

Deuteronomy 22:9-11

> You shall not sow your vineyard with two kinds of seed, lest the whole yield be forfeited to the sanctuary, the crop which you have sown and the yield of the vineyard. You shall not plow with an ox and an ass together. You shall not wear a garment of mingled stuff, wool and linen together.

The points at issue in Deuteronomy 22:9-11 are different to those of Leviticus. In Leviticus 19:19c the prohibition was against sowing 'two kinds' in 'fields'; in Deuteronomy 22:9 it is against sowing in vineyards. Also, in Leviticus 19:19b, the issue was of breeding two kinds of (probably domestic) creatures; in Deuteronomy 22:10 it is ploughing with an ox and an ass. In addition, whereas Leviticus 19:19c speaks generally of cloth made of 'two kinds (Heb. *kileayim*) of stuff', Deuteronomy 29:11 identifies a prohibited mixture as being 'wool and linen'.

Notwithstanding these differences, we would expect Deuteronomy 22:9-11 to raise the same didactic questions as Leviticus 19:19. Certainly, the similarity in content is striking. There is the same threefold prohibition of mixing 'plants', 'animals' and 'clothing'. It is also notable that Deuteronomy 22:9-11 is closely followed by a series of laws governing human sexual relations (Deut. 22:13-30). This recalls the juxtaposition of Leviticus 19:19 with Leviticus 19:20. This underlines the argument made above that both Leviticus 19:19 and Deuteronomy 22:9-11 are thematically related by the issue of 'mismating', and the laws are intended to stimulate wider reflection on the proper limits of human sexual relations, especially within the Israelite community.

Deuteronomy 22:9

This law prohibits customs such as planting vegetables in a vineyard between the rows of vines, or perhaps mixing vines and other fruit trees in the vineyard. The owner of the field loses not only the grain and vegetables sown between the vines, but also the produce of the vines themselves.

This prohibition may reflect some pragmatic or utilitarian concern; e.g. a warning against forcing the land by means of overproduction. If so, its purpose may be to curb the natural human tendency towards greed and exploitation. In its favour, this interpretation is consistent with the preceding verses. Deuteronomy 22:6-7 states:

> If you chance to come upon a bird's nest, in any tree or on the ground, with young ones or eggs and the mother sitting upon the young or upon the eggs, you shall not take the mother with the young; you shall let the mother go, but the young you may take to yourself; that it may go well with you, and that you may live long.

However, in addition, it is possible that the law refers to ancient taboos associated with alien religious practices. In particular, it may reflect a certain antipathy toward Egyptian practice. There are a number of Egyptian paintings from Eighteenth and Nineteenth Dynasty tombs showing gardens and orchards in which various types of fruit-bearing trees are growing side by side. Notably, whereas Leviticus 19:19c did not contain any motivating clause, Deuteronomy 22:9 does explain the consequence of sowing 'two kinds of seed'. This turns out to be unashamedly cultic: 'lest the whole yield be forfeited [literally, 'consecrated'] to the sanctuary, the crop which you have sown and the yield of the vineyard'.

The sense seems to be that a breach of the law would result in the whole yield of the vineyard being turned over to the Sanctuary as if it were something captured from an enemy in war (e.g. Josh. 6:18-19). It is unclear whether the yield is handed over to the Sanctuary because the food is 'sacred' and hence too dangerous for ordinary human consumption. More likely, the food is made 'holy' *because* it is handed over to the Sanctuary. Presumably, once the yield is handed over to the Sanctuary, it is eaten by the priests. Having been handed over to the Sanctuary, it has become 'consecrated' and could not be eaten by anyone else. At any rate, this cultic explanation supports the view that the purpose of the law is a cultic one: to reinforce Israel's purity and her separation from the nations. As in Leviticus 19:19, it is meant to teach the importance of not getting mixed up in any foreign religious practices or alliances. Deuteronomy 22:9 has a symbolic value as a badge of Israel's distinctiveness from the nations. The law reminds Israel, at an everyday level, not to get herself mixed up with foreign peoples or foreign practices.

Deuteronomy 22:10

Some commentators[127] think that this law is about 'being kind to animals': to plough with two animals of unequal strength is cruelty to the weaker. This may well be part of its purpose. However, given that Deuteronomy 22:9-11 is interlinked, and

---

127 Ian Cairns, *Deuteronomy: Word and Presence,* International Theological Commentary, Grand Rapids: Eerdmans, 1992.

granted that Deuteronomy 22:9 is explicitly given a cultic explanation, we probably ought to look for a cultic explanation for Deuteronomy 22:9 also.

According to the criteria listed in Leviticus 11:1-8, the ox was a 'clean' animal and the donkey was 'unclean' (although it was eaten in dire emergency, cf. 2 Kgs. 6:25). This distinction is reinforced by the dietary laws prescribed in Deuteronomy 14:1-8. Donkeys were not offered in sacrifice to YHWH (Ex. 13:13), but may have been sacrificed to some Canaanite deity. Consequently, we should see this law as being less to do with 'animal rights' and more to do with not yoking together the 'clean' and the 'unclean'. The Israelites should keep apart the clean and the unclean. Among other things, this means not marrying foreign peoples. As we have already noted, this fits very well with Paul's command in 2 Corinthians 6:14 about not being 'yoked' with unbelievers, especially the sequence of thought that culminates in v. 17: 'Therefore come out from them, and be separate from them, says the Lord, and touch nothing unclean...'. The image of 'ploughing' the field also has sexual overtones, consistent with Paul's usage in 2 Corinthians 6, and with the sexual content of the succeeding laws (Deut. 22:13-30).

Deuteronomy 22:11

The prohibition in Deuteronomy 22:11 is against wearing a fabric of mixed weave, (specifically 'linsey-woolsey').[128] This combines wool, from animals, and linen, from plants. Again, it appears to be a visual reminder to maintain strict boundaries and separation. As with Deuteronomy 22:9 and Deuteronomy 22:10, this prohibition appears to be cultic in nature. As with Deuteronomy 22:9, it may have been related to some custom practised in Egypt. The phrase 'mixed stuff' (Heb. *sha'atnez*) is an obscure foreign term that Craigie[129] claims appears to be a

---

128 Cairns, 1992, p.196 translates this fabric as 'false weaving'.
129 P. C. Craigie, *The Book of Deuteronomy,* New International Commentary on the Old Testament, Grand Rapids: Eerdmans, 1976, p. 290.

word taken over from Egyptian. (This is a theoretical possibility, although we have no example of the use of this word in Egyptian texts). During the Eighteenth Dynasty various complicated types of pattern weaves were being introduced in Egypt. It is quite likely that these may therefore have had reprehensible associations for the Israelites.

This interpretation may be supported by the juxtaposition of the following law: 'You shall make yourself tassels on the four corners of your cloak with which you cover yourself' (Deut. 22:12; cf. Num. 15:37-41), where the tassels reminded the people continually of the law of God. It is plausible to suggest that there is a binary opposition here between 'clothing associated with Egyptian practices' and 'clothing associated with Israelite practices'.

For further reading

Douglas, Mary (1966), *Purity and Danger*, Routledge and Kegan Paul: London.
The classification of Leviticus seen from an anthropological perspective.

Milgrom, Jacob (1991), *Leviticus 1–16* (Anchor Bible Commentary), Doubleday: New York.
A comprehensive examination from an orthodox Jewish perspective.

Wenham, Gordon J. (1979), *Leviticus* (New International Commentary on the Old Testament), Eerdmans: Grand Rapids.
A survey from a conservative Christian standpoint.

Wright, Christopher (1996), *Deuteronomy* (New International Bible Commentary), Eerdmans: Grand Rapids.
A lively account of the laws of Deuteronomy with some thoughtful application.

---

129 P. C. Craigie, *The Book of Deuteronomy*, New International Commentary on the Old Testament, Grand Rapids: Eerdmans, 1976, p. 290.

Appendix 5

The Working Group and Peer Reviewers

Members of the GM Foods Working Group

Dr Donald Bruce (chair)
Ethics of genetic engineering, risk assessment, patenting, societal issues.
Background in chemistry, nuclear power research, plus theology diploma.
Director of Society, Religion and Technology Project of the Church of
Scotland examining ethical issues in technology. Chaired five-year study of
ethics of non-human genetic engineering, co-editor of *Engineering
Genesis.* Author of Church of Scotland and European church policy papers
on GM food, gene patenting and cloning. A director of the John Ray
Initiative.

Professor John Bryant
Plant molecular biologist.
Professor of Molecular Biology, Exeter University, and Chair of Biological
Science Ethics Committee. Interested in plant genes, replication of DNA
and gene manipulation. Set up an undergraduate course in bioethics.
Active interest in how biblical ethics relates to genetic engineering. Chair-
man-elect of Christians in Science.

Dr Jonathan Burnside
Theology and law.
Lecturer in Criminal Law, University of Bristol. Academic lawyer and
criminologist, holds a PhD. in biblical law. Interested in applied biblical
thought to public policy. He is co-editor of '*Relational Justice: Repairing the
Breach*' and the author of '*The Status and Welfare of Immigrants: The place of
the foreigner in Biblical law and its relevance to contemporary society*', and '*The
Shekhinah Departs: Seriousness of offence in Biblical Law*'.

Dr Bob Carling
Information specialist, especially in context of electronic publishing.
After a pharmacology PhD, went into publishing as ecology editor for
Chapman Hall. Committee member of Christians in Science for 13 years.
A director of the John Ray Initiative. Interested in making others aware of
what different organisations have to say together with how the Bible's ethi-
cal guidelines apply to bioethics and environmental ethics.

Dr Peter Carruthers
Agriculture and agricultural ecology.
Newly appointed Executive Director of the John Ray Initiative in
Cheltenham. Formerly Senior Research Fellow at the Centre for Agricul-
tural Strategy at Reading University. Has worked on GM food issues with
the Agricultural Christian Fellowship of which he is also Chairman.

Dr Avice Hall
Plant diseases and ecology.
Plant pathologist and ecologist at the University of Hertfordshire based in
Hatfield. Interested in patenting, particularly in relation to the developing
countries. Chair of Natural Sciences Ethics Committee. Hon. Secretary,
British Society of Plant Pathology.

Dr Don Horrocks
Theology, project management and editorial.
Twenty-five years in business in corporate banking and management con-
sultancy. Recently completed a theological PhD at London Bible College.
He is Public Affairs Manager for the Evangelical Alliance.

Tony May
Developing country issues and ethics.
Business consultant associated with Tearfund. Has worked on GM foods
for the World Development Movement. Currently researching business
ethics for the Whitefield Institute.

Sarah Legon
Secretariat, Evangelical Alliance.

Peer Reviewers

Professor Derek Burke – regulatory and policy issues.
Former chair of UK Government's Advisory Committee on Novel Foods and Processes, former Vice Chancellor, University of East Anglia.

Dr Tim Cooper – consumer, societal and environmental aspects.
Senior Lecturer in Consumer Studies, Sheffield Hallam University and Chairman of Christian Ecology Link.

Christopher Jones – alternative and mainstream farming perspectives.
Northamptonshire farmer, Co-ordinator of Agricultural Christian Fellowship. A mainstream farmer and formerly worked for CMS in Africa. He is closely involved in issues related to UK farming, developing countries and world trade.

Dr Ray Mathias – research and development perspective, Third World research.
Head of External Relations, John Innes Centre, Norwich, plant biologist.

Rev Dr Michael Northcott – theology, ethics, and environment.
Senior Lecturer in Christian ethics and practical theology, Edinburgh University, and a leading environmental ethicist.

Ms Julie Whittaker – economics, agriculture, Third World.
Agricultural economist, Exeter University.

Appendix 6

For Further Reading

General Studies of GM Issues

L. Anderson, *Genetic Engineering, Food and Our Environment*, (Dartington: Green Books, 1999). Written from a strongly 'anti-capitalist' stance, Anderson uses GM as a lightning rod for attracting opposition to the involvement of 'big business' in agriculture and food production. His new age philosophy is also apparent.

D. Bruce, and A. Bruce (eds.), *Engineering Genesis*, (London: Earthscan Publications, 1998).
A ground breaking five-year study of a Church of Scotland expert working group, presenting pros and cons of genetic engineering in crops, food, animals and micro-organisms. Written for the secular market and acknowledged for its balanced approach, it predicted many of the problems which emerged over GM crops. Second edition due end 2001.

J. A. Bryant, J. F. Searle and L. M. Baggott-LaVelle, (eds.), *Bioethics for Scientists*, (Chichester: John Wiley and Sons, 2001).
A multi-authored book which covers a range of bioethical issues, including GM crops and food. Not written from a specifically Christian standpoint although several authors analyse the various religious positions in relation to the subject of their chapters.

Celia Deane-Drummond and Bronislaw Szerszynski (with Robin Grove-White), *The Reordering of Nature: Theology and the New Genetics*, (Edinburgh: T. & T. Clark, 2002).
A series of chapters from various theological standpoints on many aspects of genetic engineering, with some challenging insights.

M.J. Reiss and R. Straughan, *Improving Nature? The Science and Ethics of Genetic Engineering*, (Cambridge: Cambridge University Press, 1996).
The first major book to appear on the general issues of GM, giving a readable and balanced ethical assessment. Written in the period before the GM crisis, and inclined to be positive towards the technology. One of the authors writes from a Christian standpoint.

J. Rissler and M. Mellon, *The Ecological Risks of Engineered Crops*, (Cambridge, Mass: MIT Press, 1996).
Written by US ecologists who acknowledge there is nothing unsafe in GM per se but are very strong on the precautionary principle in relation to the application of GM.

Publications of Relevant Organisations

House of Lords, *Science and Society*, Report of the House of Lords Select Committee on Science and Technology, (London: HMSO, 2000).
A landmark study belatedly acknowledging much that had gone wrong in the GM issue as well as how science and scientists have related to wider civil society.

Various **official regulatory and advisory bodies** have useful information on their websites:
The Advisory Committee on Releases to the Environment (background to the crop trials)
The Food Standards Agency

The Agriculture and Environment Biotechnology Commission
English Nature
Scottish Natural Heritage

The **Biotechnology and Biological Sciences Research Council** and the various UK crop research institutes all produce publications about GM issues and the research that is going on.

Many **green NGOs** and organic organisations have publications on GM issues, including :
Green Alliance
Gene Watch
Genetics Forum
Friends of the Earth
Greenpeace
Royal Society for the Protection of Birds
The Soil Association

Some Studies and Statements by UK Churches and Christian Organisations

Christian Aid, *Selling Suicide,* (London: Christian Aid, 1999).

Christian Ecology, *The Church of England's View on Genetically Modified Organisms – A Response by Christian Ecology Link,* (Harrogate: Christian Ecology Link, 1999).

Church of England Board of Social Responsibility, *Genetically Modified Organisms – A Briefing Paper,* (London: Church of England, 1999).

Church of England Ethical Investment Advisory Group, *Genetically Modified Organisms,* (London: Church of England, 2000).

Church of Scotland, Reports to the General Assembly and Deliverances of the General Assembly 1999, *The Society, Religion and Technology Project report on Genetically Modified Food*, pp. 20/93-20/103, and Board of National Mission Deliverances 42-45, p. 20/4.

Methodist Church : *Discussion pack on Genetics*, (Methodist Publishing House, 1999).

John Ray Initiative, *Genetically Modified Crops*, Briefing Paper No.5, *Christians and Genetic Manipulation – Are we 'Playing God'?*, Briefing Paper No. 6, (Cheltenham: John Ray Initiative, 2001).

URC : D. M. Bruce, 'Genetic Engineering – the Missing Values', *Reform*, Nov. 2000, pp.16-17; T. Cooper, 'GM Food Stop (…or Go)?', ibid., p.20, December 2000; J. Biggs, 'GM Food (Stop) …or Go?', ibid., p.20, December 2000; K. Bundell, 'Feed the World?', ibid., p.12, January 2001; D. Burke, 'A Window of Opportunity', ibid., p.13, January 2001; N. Messer, 'xxxx', ibid., pp.24-25, February 2001.

A series of short articles in the journal *Reform* of the United Reformed Church which give snapshots of some of the different views within the Christian community over GM crops and foods.